Weed the People

From Founding Fiber to Forbidden Fruit

Jeremy Daw, J.D.

Cannabis Now Publishing

ISBN: 1938567919
ISBN-13: 978-1-938567-91-9

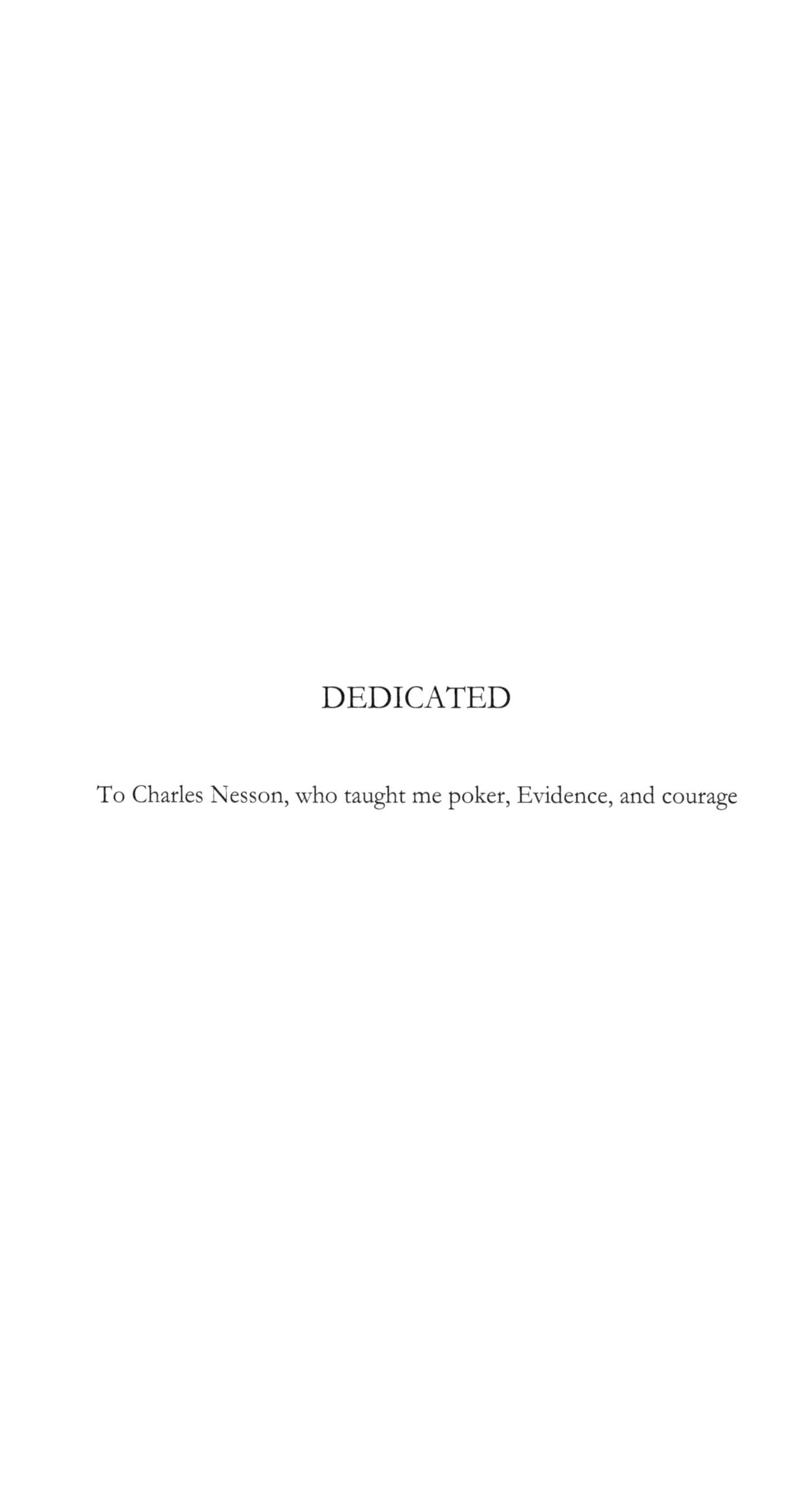

DEDICATED

To Charles Nesson, who taught me poker, Evidence, and courage

CONTENTS

ACKNOWLEDGMENTS

No one ever wrote a book alone. In addition to Jay Wiseman and Matt Whatley, who gave invaluable insight into the publishing world; Anthony DiFranco, whose tireless efforts gave the later drafts a much needed finesse; Bob Verini, who gave some early and much-needed guidance on structure; Steve DeAngelo, who offered titillating tidbits from his encyclopedic knowledge of the subject; Eugenio Garcia, who filled an instrumental role in the work's early evolution; Jeanette Mena, who offered an essential Mexican perspective; Erika Cheung, who provided research support at the project's genesis; Shiva Alcheman and Kala Snowflower, who endured multiple drafts and gave helpful commentary; Adam Safron and Randy Daw, who spared the reader many a middling turn of phrase; and the heroic staff at Sonoma Coffee Café in Berkeley, who brewed about a gigaliter of *Coffea Arabica*; the author himself made occasional edits, with mixed results.

PROLOGUE: THE CAMP STALKER

"What is a weed? A plant whose virtues have yet to be discovered."

-Ralph Waldo Emerson

In the dim predawn of history, humanity clung to life by its fingernails. On the harsh upper steppes of central Asia, between the forbidding Himalayas and Hindu Kush lurking ominously to the south and the demoralizing Gobi Desert reaching far across the north, the winter winds scoured the plains' scrubby brush and the summer sun beat mercilessly upon the cracked, parched, granular soil, defying life to grow.

Amongst this desolation, ancient human ancestors eked out their existence where to live meant to follow - to constantly chase the herds of now-extinct cattle which circled the landscape during their seasonal migrations. In the long dry seasons, the herd knew where to find water. Their broad, trailing swath left behind nutrient-dense stool to enrich the soil, enabling edible plants to thrive. Birds, too, followed the herds, bringing seeds and fertilizer from afar and building the biodiversity necessary to support human life. Yet the biggest benefits were reaped directly from the herds themselves, when the occasional sick or old member fell prey to the stone-tipped spears of its human hunters; to provide meat for food, hides for warmth and shelter, horns for signaling. On the wind-swept bare scratches of earth north of the spine of the world, such luxuries divided the lucky from the dead.

The most successful groups formed clans and tribes and sank easily into long traditions, long gone. Language formed, and loyalties, and cultures. The humans who followed the migratory patterns of prehistoric kine developed patterns of their own, stopping at seasonal camps in their long wanderings - and inevitably, the followers were followed.

No one knows how it first happened. It may have been a bird; the plant's fruit has seemingly always had a special appeal for birds. It may have been an adventurous human, driven to trial of an unknown morsel by the lashings of famished hunger - an undigested seed passing through to newly-disturbed soil. But whatever the reason, human beings first encountered the tall, hardy annual growing straight and hale beside the camp outhouses, even when all nearby vegetation wilted.

For the nomadic tribes of the central Asian steppes, its appearance was revelatory. Under its thick canopy of leaves, the earth was shaded from the harsh sun, encouraging moist and loamy soil to form where once there was dry dust. In addition to utilizing the nitrogen-rich fertilizer left by humans, the plant also reached deeply with a forked taproot to draw minerals up from the lower strata of the terrain, leaving behind a deeper, more aerated topsoil than before. Local ecosystems began to transform.

As the prehistoric tribes further plumbed the plant's uses, they must have been astonished - for they had discovered an organism seemingly engineered for human benefit. Noting its gastronomic appeal for birds, they found in the plant's fruit a reliable protein source in times when meat was scarce. Its leaves, as well, provided a ready means to add nutritious greens to the early human diet. But investigating still further, they discovered that the greatest store of uses was to be found in the clinging fibers wrapped around the stalk's woody core. From this sturdy, fibrous material, they wove wraps as warm as hides, stretched long fibers into taut bowstrings, and wrapped lines of twine into a reliable rope. With such luxuries at hand, these early tribes of men and women lived longer, fuller lives. No longer was a nomadic lifestyle of hunting and gathering needed; the tribes spread from the high steppes in the shadows of the world's tallest mountains to settle into the lush river valleys of what is now western China; and the plant followed them.

In the lands to the east, Chinese cultures called the plant "Ma." In the lands to the west, Indo-European tribes spread its name far and wide: to Assyria, where it was known as *qunnabu*; to Persia, where it was called *quonnab*; to ancient Israel, where it was named *qanneb*; and finally to ancient Greece, whose inhabitants called the prodigious herb *kannabis*.[1]

Cannabis thus became one of the earliest and most widely cultivated crops. Its use spread rapidly; by about 8,000 B.C.E., twine woven from its fibers could already be found as far as Taiwan.[2] To the early Chinese culture just beginning to cohere, the plant became so central to life that the new society called their country "the land of mulberry and hemp"[3] (the mulberry tree, botanically related to hemp, is the host species to the silk worm). Besides valuing the plant for its fiber and seeds, Chinese doctors also discovered its medical benefits very early; although the exact date of this discovery is unknown, the oldest surviving text on Chinese medicine lists a wide variety of applications from arthritis to chronic pain and proclaims the drug to be one of the most important in the physician's compendium.[4] By the birth of the earliest human civilizations, cannabis was already in wide use as a fiber, food, and medicine.

As the Indus Valley civilization spread its influence across the subcontinent of India, it quickly adopted the plant's many uses from its Chinese neighbors. By 1,000 B.C.E. its psychotomimetic resin had gained a level of social acceptance as a relaxational intoxicant - while its use was not actively encouraged, it was nonetheless recognized as preferable to alcohol.[5] From there, Aryan invasions brought the seed to Europe - which adopted the plant's fibrous but not its medical applications - and Arab traders spread it to east Africa, where its use quickly took on a spiritual and ceremonial context. Last of all, it spread to the Western hemisphere; and by the time it did so, cannabis had essentially split into two main phenotypes so distinct from one another as to spur considerable botanical debate whether the term may in fact encompass two separate species.

Cannabis sativa (named by Linnaeus in 1753) evolved in the northern latitudes of Europe, where the plant grew tall and straight, producing prodigious fibers but considerably less of the psychoactive resin which kept its flowers moist and shielded from the harsh desert suns of the south. In equatorial environments, the plant grew short, bushy, and smothered with potent resin - a specimen so apparently divergent from its European cousin that Lamarck declared it (1783) a separate species: *Cannabis indica*.[6] Before botany finally determined that cannabis is in fact only one species, the debate had already raged for centuries, seeding considerable confusion.

That confusion lies at the heart of this history and persists to the present day. Were it not for its incredible adaptability,[7] cannabis may never have had to endure the many ignominies - criminal prohibitions, public denunciations - which now plague it in nearly every country on Earth. Some nations (especially China and Canada) have worked to clear the air by drawing up laws to distinguish between the plant's psychoactive and inert forms and by allowing so-called "industrial hemp" industries to thrive as exceptions to their national anti-cannabis policies. Yet the irony of this story is that without the heritage of long-standing equivocation over the species' two most common phenotypes, those punitive policies would probably have never existed at all - at least, not for the United States of America.

Through the long lens of world history, cannabis prohibition appears as little more than a momentary blip - a brief flash to punctuate a long tale of struggle. In the U.S., the first anti-marijuana laws appeared only in the 20th century; the earliest of them will soon be a century old. Compared to the more than three centuries of American policy which not only permitted but actively encouraged the cultivation of cannabis, the span of prohibition appears slight indeed. And already there are signs that the experiment is beginning to end. But before criminal cannabis laws fade distantly into the dim mists of history - if indeed that is their fate - it is appropriate to take a step back and ask how the human species, and Americans in particular, came to live under such an ephemeral regime; in other words, how a plant so beloved for its fibers at society's founding could come to bear fruit[8] forbidden by the full force of law.

- 1 -

BACKWATER TO EMPIRE

"Make the most of the Indian hemp seed, and sow it everywhere!"

- George Washington's instructions to the head gardener at Mount Vernon, 1794[9]

When Samuel Gray heard Boston's church bells ringing furiously into the night, he assumed that the town called for help with a fire. But when he discovered that a crowd had been summoned to harry a group of British soldiers, Gray, a hardy brawler and street agitator, turned out into the streets. Others came with him, including a forty-seven year old man of mixed race named Crispus Attucks. The crowd swelled from outrage over the presence of British soldiers in the city and a series of punitive laws.

For Gray it was also personal.[10] Only three days earlier he had been at work at John Gray's ropewalk, one of the many employers providing good jobs in the colonial cannabis industry, when he saw one of the soldiers from the British garrison loafing about. Guessing his game, one of Gray's coworkers (or possibly Gray himself)[11] called at the soldier, "Do you want some work?" Many of the British garrison, lately arrived in 1768, had taken part-time work at Boston's factories, crowding native Bostonians out of the job market. John Gray's was one of the few left in town which still proudly employed Patriots and there, contempt for the soldiers ran high. The Brit apparently didn't realize this, responding with an eager yes. "Wee then, go and clean my shithouse," came the sneering reply. The soldier challenged the Bostonians, who promptly chucked him out into the snow. But the redcoat returned later that day with a dozen of his friends from the garrison, and pummeled Samuel Gray and his Patriots.

That redcoat may have been Matthew Killroy, a soldier with the 14th

Regiment under the command of Captain Thomas Preston. A witness at his trial would later testify that he heard Killroy swear that he "would never miss an opportunity, when he had one, to fire on" Bostonians. Another account recalls the ominous prediction of the wife of James McDeed, a British soldier stationed in Boston, who declared on the evening after the fray "that before Tuesday or Wednesday night [the British] would wet their swords or bayonets in New England people's blood."[12]

The mob arrived at the Statehouse, and Samuel Gray saw Killroy among the redcoats. Attucks, with Gray not far behind, led the mob to within inches of the soldiers' pointed bayonets. After a few heated moments the large black man snatched the bayonet of British soldier Hugh Montgomery, who shouted "Damn you, fire!" and discharged his musket.

Matthew Killroy saw his opportunity. After a pause he shot Samuel Gray while the ropemaker's hands were still in his pockets, blowing a hole in his head "the size of a fist".[13] Several more of the soldiers fired sporadically into the crowd, killing Crispus Attucks and mortally wounding three more. Captain Preston, who had ordered his troops to hold their fire, took advantage of the crowd's temporary shock to retreat back to the fort.

The dispute at the rope walk, a highly visible arm of Massachusetts Bay Colony's thriving cannabis industry, played only one part in a roiling fulmination all along the coast. And although cannabis played but a small role in provoking the incident, the effect the plant would have on the fight's aftermath would profoundly alter America's history: Samuel Adams helped to write *A Short Narrative of the Horrid Massacre in Boston*, an anti-British screed which placed the full blame on the soldiers. He followed up with letters to the editors of the *Boston Gazette* in which he defended Crispus Attucks, and with further publications alongside a woodcut by Paul Revere. Author Thomas Paine used the anger over the event, by then known simply as the "Boston Massacre", and further inflamed attitudes with his famous pamphlet "Common Sense" (1776), stirring up the political will to fight the king. And in the summer of that year, a convention of delegates drafted an audacious Declaration of Independence, signed it, and shipped it across the Atlantic to add to the troubles of the absentee king of an overstretched empire. All of these documents, and many more, were printed on cannabis paper, the most popular form of paper in the world from the time Johan Gutenberg printed his first Bible on it in 1456 until the 20th century.[14]

By 1774 the colonies produced so much cannabis for paper, textiles, rope, sails, food, medicine, and even "recreational use", that the surplus could be traded to France in exchange for stockpiles of guns. When one such cache was discovered in Concord, Massachusetts, British soldiers marched to confiscate it and encountered an armed militia at Lexington, where the forces exchanged the first shots of the war for America's independence. And George Washington, a cannabis farmer, led a rebel army of soldiers clad in uniforms of woven cannabis cloth,[15] soldiers fighting under the flag of what would soon be a new nation: red, white, and blue, thirteen stars, thirteen stripes, and one hundred percent cannabis.[16]

The story of *Cannabis sativa L.*[17] in the United States is as old as the country itself. Even older, in a way, because before colonists ever arrived at the first permanent English settlement in the New World at Jamestown, Virginia in 1607, they were already legally obligated to use what arable land they could find there to make hemp quotas.[18] Indeed, this very prospect was a major reason why the English Crown and the Virginia Company of London risked the considerable sum of money needed to induce and equip settlers to undertake a long, perilous ocean voyage, endure disease and confront wary native tribes, all in an attempt to colonize a flea-ridden backwater at the edge of an empire. If it weren't for cannabis, the United States of America may never have existed at all.

The events which predestined England's North American colonies began in 1530, when a conquistador named Pedro Quadrado received an *encomienda* from Carlos I to grow hemp in New Spain.[19] The Crown granted Quadrado this rare and coveted charter of entitlement to newly discovered lands on the explicit conditions that Quadrado grow as much of the crop as possible.[20] At this point in its history, the Spanish war machine busied itself like a colony of ants to outfit the navy which would assure its coming imperial dominance, and Quadrado's encomienda represented an experiment to see whether this "New World" could amply supply Spain's military needs. Preliminary results apparently pleased the king, who in 1545 ordered "the Viceroys and Governors [to] mandate the cultivation of hemp and flax[21] in [New Spain] and [to] get the Indians to apply themselves to this farming."[22] Hemp production continued virtually uninterrupted in Mexico for the next three centuries.

With hemp grown in the mountain slopes around Tenochtitlan and in the highlands of Peru, the Spanish built a fleet known as "The Invincible Armada": with 28 massive galleon warships, 123 smaller vessels, and a total of two and a half thousand cannon, it was as much a floating fortress as a fleet. No other country in Europe dreamed of challenging Spain's naval supremacy, and with this impenetrable dominance the kings of Iberia ruled the newly discovered continents virtually unchallenged. With their lucrative silver claims and annual hemp crops, the Spanish had climbed to the top of the world.

They were a model to follow. Since Columbus' first bungling expedition in 1492, all of the most powerful European kingdoms sought the raw materials to serve as the grist for their industrial mills in far-flung lands like India, Africa, South America, and the East Indies. The Portuguese sailed via the Cape of Good Hope to India, seeking textiles; the Dutch busied themselves in the East Indies cornering the market for exotic spices; and Spain and France had an early start in the Western Hemisphere, which flooded the international markets with Spanish silver and French furs.

England had nothing. It had no empire and few resources; its economy was modernizing, but the small island country was running out of space. Spain had already blazed the way with a gamble proving that the rigging to set the wind behind the prows of a global empire could be grown right out of the ground. All England had to do was copy the strategy.[23]

But a Tudor Man 'O War required a full 90 tons of hemp fiber to rig.[24] The sails and rope were woven by loom and ropewalk (a long, low factory which manufactured rope) out of the interior bast of cannabis sativa, the strongest plant fibers in the world. Those fibers were extracted by use of a hand brake, a hand-powered iron hinge which catches the tough stalks of the hardy annual plant between three blocks of wood spaced a few inches apart; if the laborer swung with enough force, the impact of the hinge broke the woody hurds which make up the stalk's core, separating them from the stringy fibers.[25] But even before braking, the cannabis had to be *retted*: cut at the base and allowed to partly decompose in dewy fields or freshwater streams, so that the sticky lining which adheres the fibers to the hurds can break down, making the fiber extraction possible.[26] And even before retting, the cannabis had to be grown, at a rate of 18 pounds of dressed fiber per acre of fertile, well-draining soil in a season;[27] so the 90 tons of fiber comprising the rope, canvas, and sail required to build one ship of the line required the entire annual output of almost nine thousand acres. It was a tall order for a small island.

One solution which didn't work was to grow the hemp at home. Elizabeth's father Henry VIII, who ironically banned hops, proclaimed in 1533 a fine of three shillings, four pence to be levied on any English farmer who did not set aside a portion of his land to grow hemp.[28] Elizabeth increased these fines to five shillings thirty years later, but neither decree met with great success; the fine was eventually repealed in 1593.[29]

Another option was to shop around. Russia had vast fields of farmland tended by a large, cheap labor force of nearly-enslaved serfs and ample freshwater streams to ret the fiber crop. Elizabeth struck a deal with the Russians, and before long, 97% of all of England's cannabis came by sea from Baltic ports[30] – but there were issues. First, the quality of fiber coming from Russia simply could not compare to that of Venice, the finest in Europe and supplier to England's Spanish rivals;[31] second, the Russians had every commercial advantage over the English, and weren't reluctant to press their advantage. Weighing down shipments with rocks and water (fiber being sold by weight), and shipping rotten fibers were common Russian shenanigans.[32] This arrangement left much to be desired on the part of the English.

Thus Elizabeth's decision to found North American colonies hinged on the blade of an uneasy bet - an attempt to beat Spain at its own game, founding New World colonies to harvest the hemp and timber necessary to build a force to be reckoned with.[33] She failed, however, to fully appreciate the full range of uses to which the cannabis plant can be directed.

It is unknown when the first indigenous slave working the hot hemp fields of Mexico first divined the psychoactive properties of the female flowering tops of her master's plants, but it must have occurred before 1550, when Mexico's governor noted that indigenous tribes had incorporated hemp into their rituals.[34] The Native American tribes which inhabited the land before the arrival of Columbus had cultivated such a rich body of herbological knowledge that New Spain's first protomedico (a kind of Surgeon General) expressed his puzzlement as to how a group of such obviously uncivilized "savages" could possess medical expertise equal to or greater than Spain's.[35] Indeed, as the first Spaniards in the New World observed, the treasures of indigenous botany were routinely taught to every member of the tribe, so that any native they encountered, even children, could explain local plants and their uses. To a culture with a medical tradition still steeped in superstition about demons afflicting the wicked with disease, the empirically-based plant science of the Mexica was astounding.

Not that indigenous medicine hewed closely to any atheistic Enlightenment philosophy. Like the blend of rationalism and religious superstition taught in the lecture halls of Spain's medical schools, the medical traditions of the Aztecs and contemporary tribes often spoke of disease in terms of a supernatural world found just on the other side of the mundane, in which shamanic *curanderos* held the power to either cure or curse, illness might be caused by demonic possession, and deities appeared to the trance-like state of the healer to reveal the nature of a patient's ailment.[36] This final element of Mexica medicine was inextricably entwined with the local flora of Mexico, which boasts the richest diversity of psychoactive plants and mushrooms in the world:[37] *curanderos* commonly ingested indigenous organisms like the peyote cactus or the psilocybin mushroom to enter an altered state of consciousness, in order to perceive the patient's malady more clearly. In the hallucinatory state caused by these psychoactives, Mexica medicine men claimed to commune with deities which actually lived inside the cactus, mushroom, or herb itself, and to merge their consciousness with the deity's omniscience, presaging the modern concept of the entheogen (literally, "the god within"). Thus, when Catholic missionaries arrived in Mexico offering sacraments which they claimed were the blood and flesh of Christ, the indigenous population readily accepted Catholic Sacrament as yet another way to commune with a god by eating him.

In time, this culture of acceptance, more than any divergence of opinion, would spell trouble for the Spanish hemp project. While *los indios* held no strong compunctions about assimilating Catholic symbolism and ritual into their preexisting culture, they tended to stubbornly resist the demands of missionaries from Europe that they give up all other gods but Christ; and indeed, the exhortations of the Spanish may have been ultimately self-defeating due to a fundamental misunderstanding of Mexica spiritualism. For example, the common admonishment of Catholic friars that the "deities" encountered by the user of peyote or sacred mushrooms were not actually gods but manifestations of the devil who appeared in an attempt to "trick" the imperiled soul might have actually reinforced indigenous beliefs, which generally regarded each deity as simultaneously embodying both godlike and demonic, benevolent and malevolent qualities. The indigenous tribes of Mexico simply didn't believe in completely abandoning one's old traditions before adopting new ones, and tended to form syncretic practices instead. Catholic bishops and scholars noted this effect with consternation, with at least one authority claiming that the more similar an indigenous religious practice was to a Catholic one, the more surely did that practice come directly from the devil.[38]

It was this religious tension that brought the regulation of drugs under the jurisdiction of the Spanish Inquisition, an ominous precedent which grimly foreshadowed future laws and policies. By 1620, the Inquisition had grown weary of trying to compete in the bazaar of sacraments, declaring instead that "from here forward no one of whatever social status can use... peyote, nor any other [plant] for the same or similar effects, under no title or color nor shall they encourage Indians or other persons to take them understanding that if they do so... we will proceed against the rebellious and disobedient... as against persons suspected of violations against the Holy Catholic faith."[39] This edict, promulgated on June 19th, was the first anti-drug law in the New World.[40]

Yet all of this remained beyond the ken of Queen Elizabeth, who determined to found colonies of her own, risking the grave consequences of provoking Philip II of Spain, his allies in the preeminent Hapsburg royal family, and his mighty Armada. Nevertheless the Queen took her chances, founding a colony at Roanoke, Virginia in 1585. The move proved disastrous. As soon as England laid bare her ambitions in North America, simmering tensions with the Hapsburg Empire began to boil over. Employing skillful politicking, the Spanish Crown began to turn the most powerful European heads of state against the wayward Elizabeth, uniting the Catholic kings against the specter of an encroaching global Protestant empire. Forces from Holland to Spain began mustering for an invasion.

The Queen would have to find a way to beat them on the water.

Backed into an untenable corner, Elizabeth risked her life and kingdom on a desperate chance, inspired by the daring antics of an independent operator. There was a bold English sailor who had managed to make himself a thorn in the side of the Armada for more than a decade, plundering treasure from captured ships and exasperating naval commanders with his daring maneuvers. His name was Francis Drake, and his ship, the Privateer, was the fastest on the seas. Through a series of intrepid raids, the swashbuckling pirate had proven that smaller, more maneuverable, less cumbersomely armed ships could prevail against the lumbering Armada and its reliance on brute force.

Elizabeth had first enlisted Drake's aid in 1577, when she asked him to lead a strenuous raid on the Spanish holdings on the western coasts of South America. This Drake accomplished, completing the second-ever circumnavigation of the earth and hitting every Spanish treasure store he could find along the way. By the time he sailed back to Plymouth his hulls held so much treasure that the Queen's 50% share exceeded the rest of the Crown's income for the entire year.

The English Navy took note of Drake's success, building dexterous ships of a new design, the race-built galleon, a marvel of marine engineering born of the insights gained by Drake's mentor John Hawkins in his travels around the world. The race-built galleon possessed far superior speed and maneuverability compared to its clumsy Spanish rivals, even in rough seas. While Drake harried the Armada in the Caribbean in an attempt to buy the English more time, the shipbuilders at Plymouth and elsewhere on England's coast kept hard at work.

Drake bought his country as much time as he could, disrupting Spanish supply lines in the Caribbean and Mexico and keeping valuable warships tied up chasing him all over the seven seas. But there was only so much a gaggle of sleek ships could do; by 1587 the Spanish were ready to attack, and England still had no fleet to match the Armada.

Gathering a massive army of 18,000 Spanish soldiers and 30,000 Dutchmen, the Hapsburgs made ready to launch a full-scale assault. If they had managed to land on English shores, Elizabeth's life would be forfeit, her country would be subjugated, her New World plans would die in the cradle, and this book would have been written in Spanish. The dry docks at Southampton bustled with an existential fury.

At the time, no one could have predicted the outcome of the desperate English gambit. Drake, in one of the great virtuoso performances of his career, put off the invasion by another six months by, burning, disabling, or capturing one hundred Spanish ships in a single exhausting raid in the Bay of Cádiz. But the Armada bounced back faster than Drake anticipated, and when it finally launched from Lisbon in late May of 1588, the enormous fleet took two full days to leave port. When the Spanish were spotted off the coast of Cornwall on July 19th, the English had to scramble to ready their ships, mustering every sleek, maneuverable craft they could launch, with Drake's *Privateer* on point. Even so, the prospects looked grim, as the relatively light armament of the English ships lacked the power to reliably penetrate the sturdy hulls of the Spanish ships. What's worse, the English had to contend with another foe which put their hemp sails and rigging to a severe test: fierce gales from the North Atlantic, fed by unprecedented summer storms.

But the winds which the English sailors cursed at first were destined to save them, because the clumsy Spanish floating fortress fared much worse; by the time the Armada sailed into the English Channel, five of their largest ships had already been forced to return to Spain, defeated by the winds. Once in the Channel, the high, lumbering hulks of their ships strained just to maintain position, driven ever eastward by the fierce prevailing gales beating against them, while the Channel got narrower and narrower, making their maneuvers more difficult. By contrast, the low-to-the-water ships of the English navy, with Drake's Privateer on point, maneuvered ably in the hostile winds. Then, in the moment which upset every axiom of naval power to that day, Drake shouldered the fate of a kingdom on the strength of his cannabis.[41]

At the battle's beginning, the English Navy held position to the Armada's east, attempting to bar Spanish entry to Calais, where the Hapsburg army waited to board. But as the storms pushed the English fleet eastward into narrower waters, what had seemed a prudent position on calmer seas had begun to resemble a mistake, and the English ships crowded ever closer together, leaving little room to evade the cannon fire of the Spanish. Drake decided to try to turn the tide. Under the cover of night and just to the Armada's side, the English navy tilted their hardy sails to tack directly into the withering wind. This maneuver required the English ships to rapidly shift their rigging back and forth to zig-zag their way directly into the face of the storm, placing enough strain on their sails and rope to founder any ship rigged with inferior fiber. Yet the hempen sails of the English navy survived the move even in the midst of the Channel's fiercest winds on record, and by morning Drake sailed straight at the Spanish with all the swiftness of the storm at his back.

That day, the English sailed circles around the Armada, which had to abandon one of its most valuable ships, the *San Salvador*, after it lost control and crashed into the *Rosario*, a smaller ship in the same fleet. Hope glimmered in England yet.

When the *Privateer* conducted a nighttime raid of the San Salvador to steal badly needed gunpowder, Drake's hopes rose even further; there in the hull of the wrecked ship, the pirate discovered that the Spanish had committed a critical error. The San Salvador, he discovered, had been packed so tightly with needed provisions for the land invasion that the crew had no room to reload their cannon! If other ships in the Armada had been packed the same way, the dreaded fire of the fleet may soon be exhausted. The English redoubled their attack, darting around the Spanish to draw their fire.

It was too late. Before Drake could come at his opponents in earnest, the Spanish managed to reach their goal: the harbor at Calais. There, sheltered from the wrath of the seas, the Armada licked its wounds and began loading the army intended for the destruction of their foes, camped just over twenty miles distant.

That night the English played their final card. Loading up a handful of ships with tar, pitch, and gunpowder, they sailed them by the hand of a skeleton crew directly at the harbored Armada and, when nearly at the harbor, lit them ablaze. Seeing the fiery maelstroms coming straight for their position, the Spanish sailors panicked, cutting anchor and scrambling out of the way. Fortunately for England, the only place to go was back into the Channel, where the once mighty Armada inexorably drifted east, cut off from Calais by the nimbler English ships and inexorably overcome by the winds.[42]

Thus, while the English guns failed to cause any major damage to the Spanish ships, their sleek galleons rigged with carefully made cannabis sail and rope kept their poise while the top-heavy Spanish ships sustained heavy damage, and the remnant of the *Armada* was forced to retreat, defeated more by the weather than the opposition. Of 130 ships sent to England, 50 never returned to Spain. A devastated King Philip despaired, "I sent my *Armada* against men, not God's winds and waves." In one of history's most dramatic reversals of fortune, Queen Elizabeth I of England found herself in the fall of 1588 at the head of the the world's preeminent naval power. England could finally create the global empire its Queen envisioned. And this time, there was no *Armada* to deter them.

Elizabeth herself did not live to see the establishment of the first permanent English colony in Virginia; that feat was accomplished by her successor, James I, whose policies emphasized the importance of cannabis as much as Elizabeth's. Although the new colony - Jamestown - was named after the king, it was actually financed and controlled by a group of investors called the Virginia Company of London. A joint stock company, the Virginia Co. was a forerunner to today's corporations, enjoying limited liability and many of the other advantages granted by modern corporate law in exchange for financing the Crown's ventures in North America. And because Jamestown was a for-profit venture, the colonists involved had to follow company policy, which required each colonist to grow a quota of hemp.[43] But despite their contractual obligations, the American colonies apparently didn't produce enough for the king, who made cannabis production mandatory in every English colony in 1611.[44] The Virginia Company, not to be outdone, increased the number of required plants to 100 per colonist in 1619.[45] The colonial governor, leading by example, pledged to grow no fewer than 5,000.[46] Cannabis grown in America was of high quality,[47] and when fed into England's advanced manufacturing system, Virginian hemp helped England to build the most powerful navy in the world. For the Virginia Company, cannabis simply meant reliable returns: the stable income provided by the English Navy's steady demand kept the colonies profitable while all other ventures failed.[48]

A strong colonial economy meant a powerful imperial navy, and throughout the 17th century American hemp powered the pursuit of England's rivals all along the eastern seaboard of North America. On May 17th, 1655 the Spanish had to quickly evacuate a strategic Caribbean island, another casualty of Spain's long inexorable loss of territory to nimbler naval power. Any Spaniards who couldn't make it to the docks in time to escape the onslaught of William Penn had to retreat to the hills in the island's center, doing their best to carry on the resistance from there. But this small group of resistance fighters hid on the wrong side of geopolitics; Spain had lost yet another prize to England.

The prize was Jamaica, a rich green jewel enjoying abundant tropical sun and situated strategically to control the Caribbean trade. The Spaniards conceded, but as a parting gift released a group of slaves, armed them, and encouraged them to agitate against their new English foes. They obliged. In time these freed slaves and Spanish resistance fighters formed a homogeneous group known as the maroons, camping in the highland hinterlands, providing refuge for runaway slaves, and annoying British overseers. Throughout the 18th century, the British governors tried and failed to put down the rebellion, and the maroons retain a measure of autonomy even to the present day.

In time this rebel counterculture brewing in the Caribbean would spell great trouble for white society to the north, but at the time the insurrection paled inconsequentially compared to another development newly arrived to the Colonies from balmier seas. When John Rolfe, the English gentleman who would eventually marry Pocahontas, brought a sweeter cultivar of *nicotiana* from Bermuda and planted it in Virginia on a trial basis, the results astounded everyone. Suddenly Englishmen the world over went wild for Virginia tobacco, paying top dollar for its fine flavor and smooth smoke (not to mention its addictive properties). Had the Virginian colonists not been required by law and contract to grow a minimum amount of cannabis, they might have devoted all of their lands to growing this now more lucrative crop - which would have been disastrous. In fact, many of the colonists did plant as much tobacco as they could, hoping to emulate Rolfe's commercial success - so many, in fact, that the price for tobacco bottomed out several times, dragged downward by sudden gluts in supply. The stable returns provided by cannabis repeatedly saved the colony from bankruptcy, when the drops in the price of tobacco caused by this overshoot of production would have otherwise consigned the project to financial ruin.

By 1650, the population of Virginia Colony had swelled to about 30,000, and had spread far beyond Jamestown, which was abandoned in 1699. By mid-century, European settlers had established a presence from New England down to Georgia, and a steady stream of new colonists arrived to work the plantations, from many backgrounds. Some craftsmen came from countries such as Germany and Poland to build the colonies' self-sufficiency; in 1619, the Virginia Company sponsored the passage of a group from Scandinavia and Poland specifically for the local processing of cannabis fiber.[49] Nor were all of the immigrants willing Europeans. The first Africans in the New World arrived the very same year and were considered indentured servants rather than slaves and many managed to work out their freedom - a few exceptional cases, like Anthony Johnson, even became rich in their own right. Soon enough, however, the laws of the colonies changed along racial lines: indentured servitude, an institution that had oppressed both white and black Americans, soon gave way to the "peculiar institution" of slavery (legally codified in 1654), overwhelmingly reserved for Africans and Americans of African descent.

Using land conquered from native tribes, the Virginia Company grew tobacco from Bermuda and cannabis from Europe using the labor of slaves imported mostly from West Africa – but kept all the profits for itself. King James, appreciating the success of the colonial experiment, revoked the Virginia Company's charter and decreed Jamestown Colony to be a Royal Colony in 1624, because kings can get away with things like that. England's prospects in the New World looked strong, but even so early in colonial history, subtle events had begun to lay the foundation of independence.

The first American shipbuilding facility, modeled after the successful English practices, was established in Plymouth, Massachusetts in 1629, serving as the start of an American rival to the mighty English merchant navy. Shortly after came the first American rope-walk in nearby Salem, in 1635.[50] Soon the colonies were trading their surpluses abroad, even to the Dutch, another English rival. Parliament responded with restrictive trade legislation culminating in the Navigation Act, forbidding the English colonies from trading with any country save England, and from shipping any goods except on English ships.[51] But the English were defeated by their own aggressive expansion; at a time when almost all English ships headed for India instead of America, these laws simply made shipping at all nearly impossible, and only increased the economic independence of the Colonies.[52]

In the meantime, America went nuts for hemp. Not only shipbuilding, but also textile manufacture exploded. By 1630, cannabis had become the source of half of the Colonies' winter clothes and almost all of their clothes for summer.[53] Demand was on the rise and imports could not keep pace, so the Colonies began passing laws to increase cannabis production. In 1637, the brand-new General Assembly of Connecticut mandated that every family sow a teaspoon of cannabis seed each;[54] and Massachusetts passed a similar law in 1639, so this strategy was apparently successful. The fledgling city of Boston, determined not to be left behind by the explosion of industry, sponsored an English rope maker named John Harrison to set up shop there in 1642,[55] the beginning of a burgeoning of new cannabis industry all along the coast. Other colonial governments tried the innovative approach of encouraging cannabis production by accepting cannabis seed as payment of taxes, including Virginia (1682), Maryland (1683), Pennsylvania (1706), and Massachusetts (1735). In 1722, Connecticut set a guaranteed premium price for cannabis fiber, while in 1762 Virginia maintained the tried-if-not-true practice of fining those who produced too little.[56] Even the English, desperate to get a piece of the cannabis profits from sales up and down the coast, offered, in 1662, to trade two pounds of fine cured tobacco for every pound of unprocessed cannabis fiber from the colonies. Few colonists took them up on it.[57] Why should they? All the tobacco came from America anyway, and the Colonies needed their cannabis.[58] Boston, again leading the way in 1718, imported a small army of professional spinners and weavers from Ireland,[59] just to keep manufacturing on pace with fiber production and demand.

Inevitably, early Americans discovered the same extracurricular thrills of the crop's intoxicating flowers as did Mexican indigenous peasants and Jamaican slaves. As many have noted,[60] both George Washington and Thomas Jefferson set aside significant portions of their lands to grow cannabis - and not just for cloth, as will be seen. Benjamin Franklin built his fortune in large part by becoming America's top paper producer,[61] but not by cutting down trees – rather, he, like almost everyone else,[62] used cannabis, which provided a much more durable product than wood pulp[63] and required a fraction of the acreage.[64] Jefferson filed the first U.S. patent for a horse-powered cannabis fiber separator,[65] apparently feeling sorry for his slaves,[66] who still had to use a hand brake. It's understandable why Jefferson, who left his slaves their freedom in his last will (apparently motivated by compassion)[67] would feel a twinge of guilt ordering them to perform "one of the hardest jobs known to man."[68]

And then there was the case of George Washington, who had cannabis seeds special-ordered from India[69] - a curious fact, given that hemp seeds were hardly scarce in Virginia.[70] In fact, the man who would become America's first President went through all the trouble and patient waiting, almost two hundred years before FedEx, to obtain seeds of a very special variety: indica.

The strains which even today still bear the name of their motherland were renowned, then as now, for their strong medicinal potency.[71] Whether or not Washington desired these strains to ease his chronic tooth pain, as some have alleged,[72] there seems to be little doubt that Washington had a keen interest in the indica's fragrant flowers. Upon receiving one shipment, Washington famously wrote to Mount Vernon's garden master, "Make the most of the Indian hemp seed, and sow it everywhere!"[73] Apparently the future president's enthusiasm got a bit the better of him, for he later lamented in his diary that he "[b]egan to separate the male and female plants… rather too late" (August 7, 1765) and that he later "pull[ed] up the [male] hemp. Was too late for the blossom hemp by three weeks or a month." (August 29, 1766).[74] Washington apparently discovered what many a marijuana grower has learned to their chagrin – that male cannabis plants can pollinate the female buds, reducing the potency of the drug.[75] In a possible attempt to avoid this contingency, Washington had his "Indian hemp" sowed down "by the watering hole," where they would apparently be protected from the pollens of his cannabis grown for fiber.[76]

Nor was he alone if Washington used part of his cannabis crop for homeopathic purposes. Thomas Jefferson admitted to smoking cannabis in his diary, and even, it has been alleged, exchanged smoking blends with Washington.[77] Unlike today, dried cannabis flowers fetched a much lower price than tobacco,[78] often considered a luxury at a time when British contracts had bought up most of the supply; indeed, it would have been quite natural for most working-class colonials to prefer the cannabis abundantly available in the fields.

But the easy freedom of careless smoke rings rising from hemp fields changed almost overnight in 1760, when England finally succeeded in subduing India. The British Empire turned her full attention to America. A year later there were English ships at every port, demanding the raw materials long ago promised. And not just merchant ships, either: English Men O' War soon became a common sight in American harbors, and in unruly Boston, those ships came accompanied by a permanent garrison of soldiers to enforce the laws of the king.

And what laws. While restrictions like the Navigation Act or the Wool Act of 1699, which banned importation of wool to the colonies,[79] had caused the colonists little pain, new decrees came down from the throne with alarming regularity, enforced from behind the firing lines of British musketeers: the Sugar Act (1764), the Stamp Act (1765), the Townsend

Acts (1767), and a slate of 1774 decrees affectionately known as the Intolerable Acts. Some seemed intended to punish wayward colonial behavior, but most were passed to raise money the king badly needed to pay his considerable war debts. The subjugation of India had been enormously expensive, and the fiscal situation became even more dire when the East India Company, the massive corporation chartered to oversee British interests in the subcontinent, nearly collapsed from mismanagement and turned to Parliament for a bailout.[80] Parliament had no choice but to comply – India was just too big to fail. But it was the American taxpayer who was ultimately on the hook for the bill.

The colonists, while rather reluctant to retroactively pay the wages of reckless and incompetent businessmen and the graft of a distant king's foreign war of conquest and exploitation, looked to their cannabis crop to provide the needed money, which yielded sizable surpluses in the 1770s.[81] Portions of these surpluses could be traded abroad, in defiance of the king's orders, and indeed a black market for cannabis fiber grew right along with discontent with the king's policies. The French, longtime rivals to Britain, were eager to snatch up American cannabis to build their own navy.[82] Even more crucially, in selling American cannabis in French markets, the colonists gained access to a commodity that, in light of British despotism, became increasingly valuable in their eyes: guns.[83]

And so the colonies came full circle. One of the main crops behind the audacious American experiment, intended to produce to supply the manufacturing economies of European powers with raw materials, ultimately weakened the colonies' ties to the European kingdoms by making the colonies economically self-sufficient. A main driver of the colonial economy, it provided them with efficient transportation, clothes, and pamphlets while it was traded abroad for weapons and other supplies.

In short, cannabis provided the colonies all they would need to start a revolution.[84]

- 2 -

THE CHICKEN EFFECT

"The greatest service that can be rendered to any country is to add a useful plant to its culture."

-Thomas Jefferson

When describing unexpected effects from minute changes to complex systems, many scholars refer to "the butterfly effect," after the famous 1972 lecture by Edward Lorenz on whether the flap of a butterfly's wings in Brazil could set off a tornado in Texas. Yet Lorenz might have found a more fitting antecedent in American history, when a massive disruption of global commodity prices, the near-bankruptcy of the world's largest corporation, the collapse of one of humanity's most ancient civilizations, and the worst drug addiction epidemic to ever hit the United States all resulted from the fateful choices of a chicken living outside Savannah, Georgia in 1793.

History records no name for this bird, but it is known that its life was hard. Times were hard all over the state. This bird's owner, like so many others who migrated to the South in the early heady years of independence, had sought to convert conquered land into a cash crop bonanza and found the business to be far harder than she reckoned. Tobacco, which had made men so rich in Virginia, simply couldn't grow well enough in Georgia to turn a profit; food crops generally couldn't pay for themselves; and hemp, the mainstay of the more northern economies, rapidly reverted to short, bushier forms in the abundant sun south of Kentucky,[85] yielding little useful in the way of fiber. There had been rumor, however, of a new fiber crop which might even replace venerable cannabis in the textile market, if only Southern farmers could figure out an efficient way to produce it. That fiber was cotton, which had been successfully grown in India for centuries. The owner of the plantation in question, Catherine Greene, had tried sowing Indian cotton seed, reasoning that the plant would thrive under the Georgia sun with days long and hot like those of its homeland, India. And while she eventually found a strain which grew well in the sandy Georgia soil, this success only created yet another problem: the plant's fibrous, valuable flowers tangled up with prickly dark seeds which had to be separated by hand from each and every tiny tuft of fiber. A cultivated Southern lady wasn't about to do all that work herself.

Which was, of course, where slavery came in. Like many other Americans of means and privilege, Ms. Greene brought a cohort of African slaves to Georgia, thinking that she had found the solution to all her cotton-pickin' problems in the form of forced labor. She reckoned wrong, however, since forced doesn't mean free: on the contrary, many wealthy planters went bankrupt learning that owning another human being was actually incredibly expensive. The price to acquire a slave at auction was so high as to be out of reach of all but the wealthiest farmers, as was the ongoing cost of maintaining a slave, and a human being was a long-term investment. To have any chance of recouping the considerable purchase price and turning a profit, a slave-owner had to keep his slaves housed, clothed, and fed through many harvests, many years. When a slave fell sick - a common occurrence - the slave-owner had to hire a doctor or risk losing his capital. The plantation's business internalized all the costs of keeping the slave ready to work, and given the back-breaking labor usually involved, the regular outbreaks of malaria, and the occasionally successful escape attempt, a slaveholder could only count on having so many good seasons out of a slave; and every season, good or bad, cost her money. Catherine Greene found she simply couldn't afford to have her human holdings process all the cotton she grew. Her experimental dalliances in the slave-powered cash crop economy had begun to bleed her estate dry. More than mere moral bankruptcy loomed.

Of all this, the chicken only knew that kitchen scraps had not been sufficiently forthcoming. Hunger has led many a creature to wander far, and one sultry Southern evening, the bird scratched for some grub at all the usual ground and came up famished. The desperate chicken started scouring the edges of the pen, unaware that its life was about to get much worse.

The barnyard cat was hungry, too. It didn't know what the matter was with all the humans in the manor, only that table scraps had grown scant. The other cats had hunted the field mice down to slim pickings, and after weeks of plaintive mewing to no effect, the cat was not about to hesitate in the face of opportunity. The bird waddled in the gloom as the cat stalked to within striking distance, as only cats can. When finally one wing came within reach the cat reached out and pinned the bird down with a deft swipe of its paw. The chicken could do nothing but struggle and skwawk, and the hungry feline might have thought its long-suffering patience rewarded. Unfortunately for the cat, all that it could pull through the fence was clawful after clawful of feathers.

Fatefully, the commotion caught the attention of a young man lounging lost in thought on the manor's wide porch. This gentleman, a recent transplant from New England, had had much to ponder since becoming the beneficiary of Ms. Greene's largesse; the more he learned about the tatters of Georgia's cash crop economy, the more he realized that the Greene plantation represented only one small part of a much larger economic disaster, the failing experiment of chattel slavery. While historians have debated to what extent he actually set out to revitalize the slave system in particular, most agree that he hoped to help the ailing Southern economy in general.

The kerfluffle at the fence interrupted his reverie. Straining to get a better look, the houseguest from New England saw the cat's claws pull tufts of feathers through the fence and gasped with excitement. Rushing inside like Archimedes abandoning his bath, he grabbed pencil and paper and set to work. He had solved it. While the chicken met its fate, forgotten and alone, the excited young Yankee sketched out the design which would finally make Southern slavery profitable and steal five million souls from the continent of their birth.

This was no ordinary houseguest. He had been born in Westborough, Massachusetts in a time of rising revolutionary sentiment, and as a young man distinguished himself at Yale. He thought about studying law but couldn't get the tuition together, so he booked a fateful passage to South Carolina in search of income. Known from an early age for his quick intelligence, he had been hired in advance as a private tutor for a local aristocrat, but as fate would have it he met Catherine Greene en route and gave in to her pleas to stay with her in Georgia. The Greene plantation, like so many others, had long teetered on the edge of insolvency, and after the latest round of troubles the teetering showed signs of toppling. Upon the guest's arrival, Phineas Miller, the veteran plantation manager, explained the crux of the problem: how to extract the pesky seeds from the cotton blossoms while preserving a workable profit margin? If the visitor, who seemed like a bright young man, could figure that out, the slaves could take care of the rest.

The obsession with cost control had a simple motivation. The East India Company, the largest corporation on Earth, had consolidated its control over India less than twenty years before America won her independence. With Indian cotton production maturing under the Company's guidance and exerting its effect on prices in global textile markets, America lost the advantage that had created its economic independence right at the moment its political independence was secured.

British merchants, once the mainstay of demand for American goods, found they could get all the textiles they needed using cheap low-caste Indian serf labor and even cheaper stolen Indian land, and while the finished product of cotton fabrics came nowhere close to the durability of American-grown hemp cloth, it mattered little. At the low prices in Indian markets, the British could just buy more.

In contrast to the American system of extensive investment in an owned human being, the Honourable Company's Indian business model placed a lower value on human life. If an untouchable peasant died of starvation somewhere in Bengal, what did his employer care? He had been rented, not bought, and there were many others unemployed and ready to be rented at a moment's notice. In the ruthless textile markets of the late 18th century, the slaveholding American cotton farmer couldn't compete with the high cost of low prices.

Simply put, by 1793 the East India Company was putting American cotton farmers out of business. Most plantation owners were nearly ready to call it quits, and states above the Mason-Dixon line had begun to rely on industry instead of agriculture as the major driver of their economy. Instead of low-caste Indian labor, the Northern industrialists preferred poor recent immigrants - Italians, Irish, Slavs, Jews - who flooded Ellis Island while fleeing famines, anti-semitism, or religious intolerance; yet the same basic economic principle applied in Connecticut as in Calcutta. By clinging to an outdated system of production, the Southern states fell behind not only global interests but also their neighbors and rivals to the North.

Back at the manor's drawing table, the Yankee gazed ponderously at the design he had just set down, the one which would make his name - Eli Whitney - famous around the world. His machine, which he called a "cotton gin" ("gin" is short for "engine"), mimicked the action of the cat's claw by drawing a comb down through the cotton fibers. In place of the fence, Whitney used a wire screen which allowed the fibers to pass through while keeping the seeds segregated on the other side. A simple hand crank provided all the power.

In a matter of months, Whitney had a working prototype ready for his hostess who, realizing its potential, planned to plant as much cotton as possible for the following season. Word got around, and before the young inventor could even file a patent for his device, cultivation of "King Cotton" had already spread across the South as fast as slaves could plant it.

Whitney scaled up his designs, producing versions which could be powered by horse or waterwheel and could process 50 pounds of cotton a day. Cotton's profitability skyrocketed, and thereafter anyone who could obtain enough land and a bit of slave labor could make a fortune in the South. Nearly overnight, the ROI of messily hand-processed Indian cotton plummeted relative to cheap, finely machine-processed American cotton.

Paradoxically, the machine's success also drove a massive demand for human labor, as the farmers who bought cotton plantations still needed plenty of bent backs to tend and harvest the delicate flowers. African slavery, nearly dead in North America, came roaring back.

At the same time, Whitney's invention dealt a solid blow to the hemp bonanza which previously served as a backbone of the American economy. While the market still demanded hemp fiber for rope, canvas, and paper, cotton undercut cannabis in the lucrative textile market and hemp farmers lost a strong source of financial support to a lower-quality fiber with low production costs, courtesy of a handy machine and a glut of slave labor.[86] As a single silver lining, the massive production of cotton did drive demand for hemp twine, which became the go-to material for baling together the more lucrative fiber.[87] Yet this small boon did not begin to make up what had been lost.

Indeed, the economic ripples of Whitney's invention lapped at many shores, including Mexico's. By the dawn of the 19th century production of hemp fiber around Mexico City had slumped considerably from its previous highs - but the result was a transformation, not a cessation, of cannabis cultivation, as the use of *mariguana*[88] concurrently began to sharply rise. The practice, already condemned by the Church, arrived at a moment of intense cultural argument: after it won its independence from Spain in 1821, Mexico's elites immediately began to debate the best cultural course for the new country to take. Major factions took up both sides of the question: was it better for Mexico to emulate the culture of Europe's dominant powers or to conserve its own unique heritage?

For those wishing to celebrate Mexico's unique *mestizo* (mixed race) heritage, the rich diversity of the land's plant life and the indigenous botany which described it were heirlooms to be proud of, and for a time this perspective gained ground.[89] The onset of the Age of Reason in the late 18th century sparked waves of interest in natural history among the elites of Europe and North America, spurring at least four major species-gathering expeditions into Mexico's wilderness between 1777 and 1804.[90] These expeditions, which brought thousands of previously unknown species to the attention of Europe's biologists, made a prominent European fad of the natural history of Mexico. In an exquisite irony, this renaissance in interest in Mexico's natural heritage nearly effected European cultural acceptance of cannabis on the basis of its associations with the New World - by Europeans ignorant of the fact that their ancestors brought it there from the Old World in the first place. Widespread adoption of New World intoxicants by the European elite was not without precedent (e.g., tobacco), in spite of general European disdain for the customs of the "savage", so the early 19th century represented a very real window of opportunity for the smoking of joints to become a culturally condoned practice in Mexico and even the European continent.

Yet the pettiness and hypocrisy won out, and the racist disdain for cannabis persisted. As Isaac Campos notes in his excellent history *Home Grown: Marijuana and the Origins of Mexico's War on Drugs*, the Mexican elite of the 19th century continually professed a fetish for the concept of "racial degeneracy" - the idea that immoral habits practiced to excess by a people could have not only moral consequences for the individuals but also genetic consequences for the entire race.[91] In their eagerness to present the newly-independent Mexico as a nation worthy to join the circle of "great powers" in Europe, these elites wrung their hands over the permissive miscegenation of their Spanish forbears that created a race of mestizos who, according to the theory, wavered perilously between the heritage of their enlightened European and "Oriental" indigenous ancestry. Virtuous living could propel their people into a more evolved state, but immorality could devolve los pueblos back into the forms and aspects of their indio forebears, who were believed to be descendants of Asian peoples. According to this racist theory of evolution, Asians were the lowest of the low - which made their Native American descendants only slightly better. These racist prejudices extended easily to native cultural practices.

By the mid-19th century the use of *mariguana* had come to be discussed in exactly those terms. The references to cannabis contained in the 1842 *Farmacopeia Mexicana* are revealing, containing distinctly separate entries for "cannabis indica" and the "*mariguana*" used by the indigenous people to commune with the gods.[92] While the former is described in distinctly European terms, the latter is treated as an indigenous remedy; and by that date the trend among the Mexican elite had already begun to shift to favor European-style medicine.

Mexico had been a pioneer in Western-style medicine for centuries, with the first regulations providing for the inspection of pharmacies passed as early as 1528.[93] By the mid-1600s, the Viceroyalty of New Spain had created a protomedico tribunal astonishingly similar to modern Public Health departments, responsible for the licensing of pharmacists and providing quality control. After winning its independence, Mexico updated the old colonial system by creating a new Consejo Superior de Salubridad, a bureaucratic council to regulate the distribution of drugs in all of Mexico.[94] Through this council, Mexico continued its experiments in the field of drug policy for 40 years, consolidating regulatory control of pharmacies in Mexico City and imposing penalties on apothecaries which sold dangerous drugs at a time when the U.S. and most European countries had no drug laws at all.[95]

This period in history revealed Mexico at the height of its power, when America's southern neighbor surpassed it not only in medicine but also in the breadth of its empire and the might of its army. Most of the land west of the Mississippi was Mexico; and by the 1830s white settlers from the U.S. had begun to crowd into the great Mexican state of Texas.[96] When settlers led by Sam Houston and Stephen F. Austin began to agitate against their rulers in Mexico City, the powerful general Santa Ana brought the fearsome might of the Mexican army to bear against the rebels, and it was only a stroke of luck that allowed Houston to defeat Santa Ana at the decisive Battle of San Jacinto in 1836. For nine years the new Republic of Texas clung to life under existential threat from a superior force which could ready at any time to take back by force what Mexico City had never officially conceded. After Texas' annexation to the U.S. in 1845, war broke out between Mexico and the U.S., two nearly equal rivals.

In the spring of 1847, hemp played a critical role in upending the balance. While American and Mexican armies battled from the mouth of the Rio Grande to California (spreading Mexico's powerful army thin over countless miles), General Winfield Scott prepared to end the conflict with a decisive display of naval superiority. And so, for nearly the last time in American history, the canvas and rope of the U.S. hemp industry conveyed an expeditionary force, unopposed, from New Orleans down to Veracruz, due east of Mexico City - while the bulk of the Mexican army remained overextended to the north. With an all-star cadre of enlisted officers like Robert E. Lee, Ulysses S. Grant, and Thomas "Stonewall" Jackson, Scott advanced on the Mexican capital, overwhelming the attempted defenses of Santa Ana and capturing the city on August 7th. In the ensuing treaties the once-mighty Mexican empire lost the bulk of its territory to a suddenly transcontinental United States. The powerful Mexican elite surveyed the tatters of their experiment in independence in a state of shock.

They reacted with violent cultural self-revulsion. Concerns over racial purity and racial degeneracy became much more acute, giving rise to a strong paternal impulse to shepherd the poor and lower-born away from their habits. In this hysteria for concentrating power and avoiding racial degeneracy, European-style controls took on tantamount importance.[97] Biased regulations gave advantages to European pharmaceutical firms while seeking to circumscribe the reach of the *herbolarias* dispensing native medicine[98] - an unfortunate development given that, arguably, the state of European medicine still had not reached a level of sophistication on par with Native herbalist traditions. The natural history of Mexico, once proudly celebrated, became taboo in the wake of a nationally humiliating defeat.[99]

But the ripple effects did not end there. In British-controlled Jamaica, colonial governors responded to Whitney's roiling of global textile markets by contracting a Russian agronomist and his stores of specialty hemp strains to try in the tropical climate. Apparently, the British expected the Russian strains, which could grow fifteen feet tall in a Baltic season, to produce even longer fibers under the tropical sun. But removed five thousand miles and forty degrees latitude from the place of their birth, the Russian hemp seeds did nearly the opposite.

The indica strains which George Washington special-ordered long were thought to be a separate species from the sativa strains which Europeans traditionally used for fiber, based on their apparent differences: sativas grow tall and straight, indicas short and bushy; sativas produce little medicine but strong fibers, and indicas the opposite. European botanists who studied the plant in the 18th century drew the reasonable conclusion that here were two similar but ultimately distinct species: cannabis sativa and cannabis indica (a minority of domestic scientists claimed a third species, cannabis americana).

The British experience in Jamaica flew in the face of such conventional wisdom. Their baffled agronomists discovered that, in very short order, their strains bred in temperate climes to grow tall and straight began to shorten and branch in the tropics. The bast fibers, expected to grow stronger under the abundant Caribbean sun than in the dim Baltic, actually grew weaker as the plants naturally began to devote greater proportions of their captured energy to produce large, resinous flowers. The Russian strains, though purebred sativa in lineage, began to look more and more like indicas.[100] The British soon gave up the experiment, failing to detect any virtue in the plants which they had sown.

Apparently, their slaves did not do likewise. Like the indigenous Mexica peoples who discovered the spiritual properties of *pipilzintzintlis*[101] while laboring in the hemp fields of their masters, Jamaican slaves probably recognized through cultural awareness the psychoactive properties of the cannabis plant - which in east and south Africa was known as *dagga* - after they came through unwilling passage to the Jamaican hemp fields.[102] Soon after its introduction in the British-controlled lowlands, cannabis cultivation had migrated up to the highland hideouts of the rebellious maroons, sown for a very different purpose than originally intended. Over the course of hundreds of harvests, the former land race fiber strains of the European slavers became a potent tool for mental emancipation.

As fate would have it, corporeal emancipation would not tarry far behind The watershed event occurred at Kensington Estate, Jamaica, on Christmas Day, 1831, when Samuel Sharpe, an educated Jamaican ex-slave turned Baptist preacher, held vigil over a non-violent strike as a protest against the injustice of slavery. The sit-in soon grew to such numbers that the panicked British administration sent in the troops to quell what they called a rebellion. The resulting destruction and deaths dealt the protesting slaves a sore defeat, but the incident provoked sharp debate in British

Parliament, adding new voices to the abolitionist chorus. Parliament demanded inquiries, and when the results came in the British made history with the passage of the Slavery Abolition Act of 1833, which ended slavery throughout the entire British Empire (except, conveniently, for "the Territories in the Possession of the East India Company"). It was a glorious moment, foreshadowing further waves of global freedom.

As soon as the idealists hung up their hats, the lawyers began drilling for loopholes. Freed slaves soon found themselves working the same fields as before, for low pay and subject to poor treatment, but soon they were joined by another group: indentured servants from India.[103] Sugar cane (again the mainstay Caribbean crop after the failed hemp experiment) remained as labor-intensive as any other cash crop, and the wealthy plantation owners still needed cheap labor to work the fields, so they shipped in Indian families by the boatload. Some had incurred too much debt to the East India Company; others may have simply considered a move out of India, where the economy faltered and slavery was still legal, a step up. After a set period (typically seven years), the workers were released and treated just like any other dark-skinned island resident. In time, a large Indian community grew in Jamaica, partly assimilating into the island culture.

These Indian transplants brought with them a 3,000 year old history of using cannabis for diverse purposes in a society notably permissive of the plant's intoxicating flowers. While Americans, in typical Western reductionist fashion, think of "medical marijuana" and "recreational" cannabis as distinct uses, the ancient Indian cannabis culture makes no such distinctions. Sadhus used cannabis in their religious puja ceremonies, Ayurvedic doctors prescribed it for a diverse list of ailments, and millions of people across the subcontinent used it to relax after work; but the Indian culture has long viewed cannabis holistically, recognizing that it is all of the above and none. Thousands of migrant Indians who came to the Caribbean in the wake of slavery's passing brought the practice of smoking cannabis with them - an attempt, perhaps, to forget the troubles back home.

The ruthless, careerist patrician director of the corporate ledger entry of India, a backstabbing Machiavellian named Lord Clive, shed not a tear to witness the country's mass exodus. He cared not a whit of the suffering of millions of hardworking Indian farmers who had seen their livelihoods ruined by the East India Company's obsession with monoculture, a double-down bet on cotton which left an entire country bankrupt in the wake of Whitney's invention. Clive's sole concern was profit - for his company, for his country, and for his purse. The Company's entire India strategy had proved a wash; the spurring of the Subcontinent's vast agrarian fields back to productivity became his mission - one which he would see accomplished by any means necessary.

Clive, a brilliant capitalist, found his solution by thinking outside of the box - and indeed, outside of his political jurisdiction. China lay just to the north and east; its teas, silks, and porcelains fetched handsome prices throughout England's empire;[104] but the Chinese, well aware of the mistreatment of its neighbors at the hands of European traders, refused to open any ports except Canton to trade. Even commerce in that southern port occurred only under odious regulations which resulted in massive trade deficits for any European power seeking Chinese goods. Clive might have accomplished clandestine trades beyond official notice, if Britain had any goods that the Chinese market demanded. But the famously self-sufficient Chinese already enjoyed easy access to every commodity they needed.

Every commodity save one. In 1723 the Chinese Emperor had sought the means to curb a scourge recently arrived at the Celestial Kingdom's southern ports. A barbaric race from foreign lands had begun to appear in Hong Kong, bringing many uncouth practices with them. Of these, the worst was their habitual abuse of an abhorrent drug, long known to Chinese medicine for its usefulness in moderation but also for its grave dangers when used indiscriminately. The barbarians obtained the drug from the flowers of an herbaceous annual common in India and then - this was the worst part - stuffed it into the ends of wooden pipes, lit it on fire, and inhaled the smoke while it smoldered. The Emperor's snobbish disgust became genuine concern when he learned that the habit had grown in popularity among his own people, leading to fears of an epidemic of addiction. Thus the Emperor passed an unprecedented law: he criminalized the plant. The inconvenience to Chinese doctors was mild; but the move made all the difference to Lord Clive, who realized that an opportunity had opened up to target the world's most seductive consumer market.

The Emperor's 1723 edict made the banned herb the only article in the Chinese markets which consumers could not find in the licit channels of state-sanctioned commerce, an opportunity the East India Company would move to exploit. Lord Clive planted fields of the drug all across India and persuaded Parliament to grant a monopoly on its trade in Hong Kong. Smuggling it into China through corruption and graft, his company traded with shady Chinese merchants for the exotic commodities the British had begun to crave. India's agricultural economy was radically transformed; fields of cotton were replaced with narcotics. And conflict with Chinese authorities over the smuggling soon devolved into a long series of military battles named for the drug responsible for it all:

The Opium Wars.

Clive, veteran of both war and business, organized Indian opium production under the principles of solid British business sense. Since the Portuguese, Dutch, and nearly every other European power jockeyed for a piece of the opium pie, Clive knew that the Company would have to work to outfox the competition. Seeing his opportunity, he wisely went to work on the problem of dose standardization.

Opium, after all, was easy enough to produce, but quality control had been elusive. Some opium traders cut their tar balls with poppy seeds or other herbs while others simply used mud or even dung. But Clive had the Indians follow standard, rigorous cultivation procedures, year after year, while also building up a base of factories to regiment mass production and ensure that each ball of Company opium to leave Bengal was the same as every other. The strategy worked. By 1830, company men were proudly reporting back to their superiors in Calcutta that "opium boxes bearing the Company seal were passed like bank notes among the Hong Kong traders, neither inspected nor questioned." The British trade in illegal opium exploded, roughly quintupling in just over fifteen years.

Sharks in all the world's oceans smelled the blood in the water, and along the eastern seaboard of infant America the sounds of ringing cash registers troubled sailors' sleep. Scarcely a decade had passed since the end of the British blockades at ports like Philadelphia and New York, and some daring skippers who had smuggled through the British blockade became eager to give it another go in China. Though they'd be smuggling against British interests again, they wouldn't be facing the Royal Navy. This time they'd be breaking the East India Company's monopoly on the Chinese market, secured by the Crown against all other commercial ventures anywhere in the British Empire.

The newly-independent Americans quickly appreciated that Parliament's decrees held no more sway over them than the mutterings of a maniac, and moved quickly to exploit their advantage. The East India Company, under the terms of its monopoly arrangement, could only get its opium for China from Bengal; while there was some high-grade stuff available at choice prices at ports in the Mediterranean. By 1805, old Revolutionary War smugglers James and Benjamin Wilcocks of Philadelphia had figured out how to sneak into the British Levant Company's docks at Smyrna in the Aegean, get in on the ground floor of the trade at the Turkish opium auction, and smuggle it out through the Straits of Gilbraltar, around the Horn of Africa, and across the Indian Ocean to Hong Kong, where they could make a killing by offloading on the East Asian market while thumbing their noses at the British on two continents.

So the Wilcocks loaded up their ship, the Pennsylvania, at Smyrna with three and a quarter tons of Turkish opium and sent it sailing for the Orient. It logged its port of destination as Batavia in the Dutch East Indies, but when word of the voyage reached the ears of fellow smuggler Thomas H. Perkins of Boston, Massachusetts, he could tell something was afoot. What would the Dutch do with so much opium? Perkins knew better. That ship was bound for Hong Kong, sure as sunshine.

Perkins had a point man, one Charles Cabot (also of Boston) already in Borneo. Perkins wrote to Cabot, telling him to "make a voyage to the Mediterranean to procure Opium." But Perkins wasn't done. He decided to also send his favorite nephew, John Cushing, who had just reached the age of sixteen, along with Ephraim Bumstead, a hale and dependable negotiator, who would take the young Cushing under his wing and show him a thing or two about the international smuggling trade.

The boy got quite an education. Almost as soon as they arrived, Bumstead died of cholera, leaving the teenager in charge of the whole operation. What's worse, Cushing quickly discovered that he had been beaten to market by the Wilcocks, who had dropped anchor off the Pearl River Delta, just outside Hong Kong, waiting for their target price. But Perkins needn't have worried for his nephew, who quickly replaced Bumstead with the finest mentor an opium smuggler could hope for.

The young Bostonian must have trembled slightly as he weaved through the twists and turns of the narrow alleys of Canton, delving into the black markets in the lawless depths of the city where he sought the world's richest man. His name was Houqua; nothing moved through Canton that he didn't know about, and very little that he didn't sell himself. Bereft of his intended mentor, Cushing went directly to the top of the pile to seek a new one.

No record survives of their first encounter, but the negotiations can be deduced by what happened next. Cushing begged the shadowy underlord to take him under his wing, and Houqua agreed - no one knows for certain why, but Houqua had already been deep in the illicit opium trade before Cushing arrived in Hong Kong, so he may have spotted an opportunity to continue his illegal activities through a surrogate who would serve as the perfect scapegoat if the Emperor became too suspicious: who would ever suspect the richest shadow merchant in China of teaming up with a sixteen-year old boy from Boston?

In any case, Cushing's next move could only have been engineered by Houqua. Sneaking beneath the notice of the Wilcocks fleet, the kid sailed into Canton where he unloaded his opium all at once at rock-bottom prices, flooding the market enough to ensure that prices remained depressed for weeks; then took his loose cash and empty holds all along the southern Chinese coast, picking up the finest bargains on tea and porcelain to be found anywhere in the Celestial Kingdom - via letters of introduction from Houqua which opened doors closed to all others. While Wilcocks' men remained anchored at the Pearl for months waiting for the price of opium to recover from the artificial devaluation, Cushing sailed triumphantly back to America, where his uncle became the best-connected supplier of Chinese goods on the eastern seaboard. The tables had turned.

Thomas Perkins, recognizing a good thing, returned his nephew immediately to Canton. Under Houqua's tutelage, Cushing rose quickly to prominence in the Chinese opium trade, becoming an expert at confounding the Chinese and British alike. As word of his success spread, imitators sprouted like dandelions; in time, even "respectable" businessmen like John Jacob Astor got in on the game. Astor, with the kind of certainty only a beginner could have, bought ten tons of Turkish at Smyrna and tried to broker it through the Wilcocks Company. Wilcocks, seeing an opportunity to beat Cushing at his own game, completely flooded the markets: with five tons in 1816 and with seven tons in 1817, unloaded all at once. No one could hide such a quantity of drugs, and the move came to the attention of the Emperor. When decrees came down from the Forbidden City prescribing harsh penalties for any merchant who dealt with the foreign devils, every black market player in Canton knew who had stopped the music: the next time Astor brought in a shipment of opium, no merchant in Canton would even talk to him.

It was too late anyway. All of that cheap Astor opium created a bonanza for the Chinese consumer: for years thereafter, opium prices remained low enough that any coolie could afford it, and enjoy it liberally. The seeds had been sown for an epidemic of addiction.[105]

The Chinese plight nearly mirrored a simultaneous scourge in Mexico. The invention of the cigarette, or more accurately its adaptation from the use of corn husks to smoke tobacco among the Maya, began in Mexico City at the shop owned by Antonio Charro, who stocked papers at the beginning of the 18th century.[106] From there the custom spread gradually throughout the world. Over time, the smoking of cigarettes became socially acceptable in every stratum of society, and cigarette papers became a common commodity. Unlike in most other countries, tobacco in Mexico was cheap and abundant (one benefit of being an exporter nation).

Thus tobacco consumption was only beyond the means of the lowest of the low in Mexican society: indigenous peasants, soldiers on campaign, and prisoners;[107] and in fact these very groups were primarily responsible for reintroducing mariguana into the Mexican public sphere, through the highly visible (and strikingly pungent) act of smoking. Since cigarette papers had become widely available and the cost of producing mariguana had become, for all practical purposes, nil, mariguana cigarettes presented the cheapest of the cheap thrills. And so the counterculture classic, the joint, was born.

While the natives, in close contact with the *herbolarias*, probably smoked joints primarily for their spiritual ceremonies, its popularity among soldiers may have had to do with easing the rigors of life in the barracks and on campaign, which juxtaposed brief moments of great stress with long stretches of extreme boredom. Soldiers in 19th century Mexico were held in much lower esteem than American soldiers of the 21st, and most viewed soldiers' barracks as sources of degeneracy. Yet if such was the case with barracks, it was doubly true for prisons - homes to hellish violence and wretched squalor, Mexican prisons of the 19th century were regarded as the very spawning pools of racial degeneration. There in the soul-crushing conditions documented by Mexican muckraking classics like El Periquillo by Jose Joaquin Fernandez de Lizardi or El fistol del diablo by Manuel Payno, joints may have been the only escape from the grinding misery of prison life, and at least one source notes a widespread belief among prisoners that the smoke of a joint repelled the ubiquitous bedbugs.[108]

Whatever the reason, the custom of smoking joints flourished almost exclusively among Mexico's less desirable classes, priming a culture's long-trusted drug for reintroduction under the least savory circumstances. And the contemporaneous crisis in China, about to explode into all-out war, would ensure that the United States of America would be hit with two drugs very different in action, and very similar in cultural significance.

As tensions between the East India Company and the Chinese Emperor careened toward open hostilities John Cushing, now a grown man, brazenly continued to move opium through whatever channels he could. Anchored off the Bocca Tigris, the outer mouth of the Pearl River, Cushing sold to pirates from Borneo and Malaysia, who found ways to move the opium into China through their own shadowy networks of contacts.

Houqua, fearing the Chinese Emperor, had already exited the opium market; leaving Cushing in de facto control of the Chinese opium trade. But this state of affairs would not last long. The East India Company, thus far outwitted at every turn by smaller players, decided to muscle its way back in, using the same play Cushing pioneered. After Canton was flooded with cheap Company opium, they thought the smaller players would go bankrupt, and the Company – too big, after all, to fail – would become the only game in town.

The Company failed to account for the addictive power of opium. The many gluts of cheap drugs already dumped on the Chinese market had created hordes of addicts, so that when the Company came with to flood the market again, the price for opium hardly budged at all: with so many addicts each well on their way to high levels of opium tolerance, demand was elastic, and even the large surpluses formerly sufficient to depress prices for years were instead quickly consumed. The only ripple to mark the sale of massive quantities of drugs was a steadily spreading ring of addicts.

For the Emperor, this was the last straw. Who were these barbarian upstarts who dared to defy his bans and poison his culture, undercutting its proud legacy of thousands of years as the world's preeminent civilization, reducing the world leaders in science and the arts to mere dope fiends? The Emperor decided it was time to teach these drug-peddling vermin their place.

He had no idea who he was dealing with.

The first flickerings of flame around the powder keg occurred in 1821, when one of Cushing's ships, the Emily, was at anchor around Canton's inner harbor one hot September day, late in the season, having difficulty unloading its product – the manifest read 'tin' – at a decent price. Tempers must have been running a little high. One of the Emily's crew got in a shouting match with an old Chinese woman in her little boat and threw an olive jar which hit her squarely on the head, causing the old crone to topple into the Pearl and drown.

The viceroy of Guangdong province proved zealous in his prosecution. The accused crewman was quickly apprehended, quickly tried, and quickly executed. The Chinese smugglers found to be dealing with the Emily fared even worse. The viceroy, in an unprecedented crackdown, clamped Canton shut as a steel trap.

Nothing could have been better for business. Cushing moved his operations, along with the other British and American smugglers in the game, downriver to Lintin Island, where they set up a sort of floating Costco™, offering opium on every aisle. Free of Houqua's godfathering and any other kind of scrutiny, the traders unloaded opium with abandon. Whenever the heat came snooping around from the mainland, the smugglers' elaborate lookout and alert system gave them plenty of warning to hide the incriminating evidence. Indeed, in the new paradigm of zero tolerance the only prospect terrifying enough to scare the Americans out of Canton was the possibility of the British Parliament revoking the East India Company's monopoly on Chinese opium, which Cushing and others realized would open up a flood of new competitors to drive down opium prices. As it became clear that Parliament was planning to do just that, Cushing got out of Canton in 1829, with almost all the American smugglers following soon after.

They would not see what happened next.

In the summer of 1839, a group of British sailors, it was alleged, murdered a hapless Chinese man named Lin Wei-Hei while out whoring on the Canton streets. When the Emperor's special envoy demanded the British give the sailors up, the British flatly refused.

The envoy responded with bulletins all around Guangdong bearing this stern exhortation:

> Assemble yourselves together for consultation; purchase arms and weapons; join together the stoutest of your villagers, and thus be prepared to defend yourselves. If

> any of the said foreigners be found going on shore to cause trouble, all and every one of the people are permitted to fire upon them, or to make prisoners of them.

When the British Parliament heard of this declaration, they sent in the Navy. Admiral George Elliott of the India fleet was dispatched to Canton, and when he arrived, the Chinese soldiers could only gape: at twenty-six transports holding four thousand sepoy infantry from India, sixteen men-of-war loaded to the gills with five hundred forty cannon, and four new steam-powered ships of revolutionary design, including one from the East India Company, the Nemesis. Upon arrival, he established a blockade at Canton harbor then sailed up the China coast to Amoy at the mouth of the Yangtze River, where they planned to send a written demand to the Emperor for reparations for the Company's loss of opium business during the crackdown. The message lay aboard the HMS Blonde, which ran up a white flag in sign of parley. The Chinese, ignorant of the custom, opened fire.

Elliott fired back. Sending his Indian sepoys ashore, Elliott met the Chinese army head-on – then broadsided them with cannon fire from the sea. Chusan City was destroyed by grapeshot, fired indiscriminately into crowds of men, women, and children. Nine minutes was all it took to reduce the entire city of Tinghai to bloody corpses heaped on smoldering rubble. The Emperor decided it was time to negotiate, sending a man named Secretary Kishen to deal with Elliott. Kishen did the best he could, promising Hong Kong and six million Spanish dollars as reparations for lost business. But when the Emperor heard the terms, he went into a rage. Ordering Kishen back to Beijing in chains, he sent two armies to Canton to expel the British.

Elliott, learning of this plan, destroyed Canton's defenses in advance and comfortably awaited his foes, who surrendered on arrival. The Emperor acceded to the terms of the Treaty of Nanking: Hong Kong turned over to the British Crown for 150 years; the ports of Canton, Amoy, Foochow, Ningpo, and Shanghai forcibly opened to British trade; nine million Spanish dollars in reparations for lost business, and an additional twelve million to cover the costs of the British invasion.

The proud Celestial Kingdom was brought to its knees. China's once-legendary stores of gold and silver was depleted by these exorbitant reparations, and inflation soared. With faith in the Emperor shaken, the country descended into a civil war which caused the death thirty million. When, after all this, the next Emperor made a feeble attempt to challenge the British a second time, the Nemesis, with its shallow draught and superior firepower, sailed up the Grand Canal to Beijing, where the British burned the Imperial Palace to the ground. Then Chinese capitulation was total.

With the East India Company in charge, opium quickly became legal in China again - which, ironically, destroyed the Company's profit margin once and for all. Without the risks of criminal penalties driving up the price, opium became just like any other commodity, trading at a price too modest to support the Company's bloated operations. The smaller players simply left the market. Smuggling opium into China no longer offered them any profitable return.

They weren't the only ones who left. When the countryside fell under the shadow of chaos and war, many Chinese left their homeland to seek better opportunities abroad - and some found it in the United States, which had begun an ambitious project to connect its coasts by railroad.
So when the game of global dominoes set off by a Georgian bucolic scuffle had found its every terminus, the final pieces came to rest where the first had originally tipped: a young United States. Chinese immigrants, seeking better lives, brought their addictive chains forged by ruthless Lords who reacted to the threat of American innovation; Indian indentured servants and African freemen, nestled a day's sailing south of American ports, gathered together around nocturnal fires and passed around a familiar sacrament; and a once dominant southern neighbor, newly humbled, had discovered in a drug criminalization program the apotheosis of its shame. In the following decades each of these groups would reintroduce a drug which, in an ironic twist, had been an American staple for decades.

- 3 -

THE OLD FAMILIAR STING

Sherlock Holmes took his bottle from the corner of the mantelpiece and his hypodermic syringe from its neat morocco case. With his long, white, nervous fingers he adjusted the delicate needle, and rolled back his left shirt-cuff. For some little time his eyes rested thoughtfully upon the sinewy forearm and wrist all dotted and scarred with innumerable puncture-marks. Finally he thrust the sharp point home, pressed down the tiny piston, and sank back into the velvet-lined arm-chair with a long sigh of satisfaction.

Three times a day for many months I [Watson] had witnessed this performance, but custom had not reconciled my mind to it… I suddenly felt that I could hold out no longer.

"Which is it today?" I asked, "morphine or cocaine?"

He raised his eyes languidly from the old black-letter volume which he had opened. "It is cocaine," he said, "a seven-percent solution. Would you care to try it?"

-- From The Sign of Four, by A. Conan Doyle

His legs wracked. His hip muscles contracted like spines. Chilled to the bone but still sweating uncontrollably, Samuel Taylor Coleridge awoke to agony. As he turned to his side, the nausea set in, incapacitating the celebrated poet in a vise of misery. Desperate, he reached for the drawer which contained a special medicine, bhang, brought by boat from Cairo, which could bring relief from some of his worst symptoms: nausea, hydrophobia, and wrenching, bone-bending pain.

His affliction was opioid withdrawal, and he suffered on the leading edge of an English epidemic.[109] Some twenty years later Thomas DeQuincey would publish his autobiographical Confessions of an Opium

Eater (1821), ostensibly a cautionary tale about the dangers of addiction and a popular success which secured his literary fame.[110] Addiction rates had been growing on both sides of the Atlantic for centuries;[111] as smugglers like John Cushing gave up the China trade, many of them found they could profitably unload their Turkish product much closer to home, and without breaking any laws -- circa 1850, the U.S. hadn't banned any drugs at all. And demand was high.[112] In the crowded cities, where pollution accumulated and disease spread, the masses frequently suffered painful inflammatory diseases like catarrh and tuberculosis. Opium relieved pain in the absence of more specific treatments that medicine would later develop, and also offered an escape from the general misery of city life.[113] The horrors of the Civil War caused demand to explode;[114] the Union Army imported 2.8 million ounces of raw opium and over half a million pills between 1860 and 1865. Amputation was routinely used in cases of gunshot wounds to the limbs, and the many veterans living with the chronic pain from phantom limbs and embedded shrapnel continued to use opium to dull their daily dolors long after the smoke of the war's final shots had wafted away on the wind.[115]

As the most popular oasis in a vast desert of suffering, opium came to be known as "God's own medicine."[116] Nonetheless, its reception was double-edged. It came in many forms - such as laudanum (an alcohol-based extract of opium poppy, often added to wine or whiskey), or a bolus (a large pill of poppy tar taken with water) - but to smoke it in a Chinese-style water pipe would typically provoke scandalous outrage among white society because of the practice's perceived association with a "degenerate race".[117] The most popular of all postbellum administration routes was a marriage between two 19th century medical breakthroughs: the invention of the modern hypodermic syringe and the successful extraction of pure morphine from the dark, plastic, tar-like sap of the opium poppy. Once isolated from the many other opioids found in traditionally processed opium, injected morphine became many times more potent - in its ability to dull pain, in its addictiveness, and in its potential to cause fatal overdose. The American medical academy went crazy for it.

Even if more doctors had understood morphine's dangers, the drug laws of America (or, rather, the lack thereof) would have undermined their warnings. In the late 19th century, anyone could purchase a vial of morphine over the counter, complete with its own syringe. Nothing stopped any "patient" from self-administering out of cheap supplies readily available from any corner drugstore. The spikes in usage among Civil War veterans and demure housewives seeking an escape from a nasty, chauvinistic world introduced the country to the phenomenon of addiction, and any habitual user who tried to quit became intimately familiar with the famously hellish horrors of opiate withdrawal, among the worst symptoms known to medicine.

The tragic epidemics of human suffering and misery combined with the easy availability of potent, addictive painkillers to produce a spike in use,[118] as domestic consumption grew from under 1 gram per capita in 1840 to almost 3.5 grams per capita in 1890. Between two thirds and three quarters of the growth was driven by white, middle class women, who were often prescribed opium to treat such dubious diagnoses as nymphomania, masturbation, or "sexual apathy".[119] Civil War veterans comprised much of the remainder, whether due to the chronic pain from poorly-healed wounds, the maintenance of a painkiller addiction first acquired on the battlefield, or simply a means of emotional escape from the memories of the traumatic stress of the battlefield. Old soldiers attempting to drown their sorrows in wine, whiskey, or grain alcohol might easily find laudanum in their drinks, or add it on their own, making their form of opium addiction especially complicated and insidious.

These developments greatly troubled the vanguards of Protestant culture, which had made the highest moral virtue of hard work and frowned upon any "artificial Paradise." In the first half of the 19th century, a new social movement began to appear as a result of people organizing around the idea that the use of any mind-altering substances went against God's will. Very gradually this group began to alter the terms of the national debate on drugs, transforming a medical issue into a moral one.[120]

Alcohol became an early object of contention, after the nascent "Temperance movement", as it came to be known, began to expose the ugly underbelly of urban American society and to portray alcohol as the source of these new evils. As the rural yeoman farmers of the early Republic left the countryside to become wage slaves in the cities, the drunken violence frustrated husbands could wreak on their helpless families seldom escaped the notice of those on the other side of thin tenement walls. The Temperance faction grew in response.

Typical of the Temperance pioneers was Daniel Ludlow, a devoted Presbyterian minister from upstate New York who was among the first to exhort his congregants to abstain from demon rum. Ironically, if not hypocritically, the devout Mr. Ludlow "was accustomed to drink freely of spiritous liquors,"[121] as noted by a close family member.

Ludlow's son Henry grew up sharing his father's Temperance mindset, indoctrinated not only against alcohol but also against the far more controversial subject of the day, slavery. By his early manhood Henry Ludlow made good on his Protestant Work Ethic by earning admission to study law at Princeton.[122] However, a sudden religious experience diverted Henry and his convictions into a career in the clergy and a pulpit in New Haven, Connecticut, from which he preached both Temperance and Abolition.[123] There, Henry often stirred up controversy and courted danger. He advised on the Amistad case and received a beating from a mob incensed by his radical views on race. Then, in the September of 1836, Henry's frail and chronically-ill[124] wife Abby gave birth to a son, christened Fitz Hugh Ludlow and destined to forge the foundations of a perennial American subculture.

From an early age, Fitz Hugh showed a sensitive side. When he was only four, his home became a safe house along the Underground Railroad,[125] the fugitive slave support network pioneered by figures like Harriet Tubman (who often drugged crying babies with opium to avoid detection). Fitz Hugh never forgot the first slave to shelter with his family, a man named Isidore Smith with a network of lash scars crossing his back which brought the young Fitz Hugh to tears.[126] The young boy would soon suffer even more: when his mother died in 1849, Fitz Hugh was only 13.[127]

As the child grew into adolescence his sensitivity led him into rebelliousness, causing his strict, religious father much grief. In a radical attempt to recast his son's personality in the mold of a devout Protestant, Henry Ludlow laid the guilt on thick, sending the teen to a reformatory school in Poughkeepsie where he was taught "there was a place for bad boys in the world to come."[128] The bright Fitz Hugh excelled academically at the school (and, later, at Seminary[129]), but by his late teens he had begun to question the foundations of his father's faith, and consequently his psyche began cracking under the pressure of his guilt.

Perhaps it was because of the boy's chronically frail health, which would have made visits to the doctor routine, but by this time Fitz Hugh had already become well-acquainted with the local pharmacist, a Mr. Anderson of Poughkeepsie.[130] At age seventeen and far from the strict supervision of home, Fitz Hugh began indiscriminately sampling Mr. Anderson's drugs. Fitz Hugh would later recall:

> "[H]ere especially, with a disregard to my own safety... have I made upon myself the trial of the effects of every strange drug and chemical which the laboratory could produce. Now with the chloroform bottle beneath my nose have I set myself careering upon the wings of a thrilling and accelerating life, until I had just enough power remaining to restore the liquid to its place upon the shelf, and sink back into the enjoyment of the delicious apathy which lasted through the few succeeding moments. Now ether was substituted for chloroform, and the difference of their phenomena noted, and some other exhilarant, in the form of an opiate or stimulant, was the instrument of my experiments, until I had run through the gamut of queer agents within my reach... When the circuit of all the

> accessible tests was completed, I ceased experimenting, and sat down like a pharmaceutical Alexander, with no more drug worlds to conquer.[131]

Fitz Hugh conducted these experiments with the full complicity of the pharmacist, who broke no laws in giving all of these drugs to a curious teenager.[132] In part, this was because of legitimate concerns about the reach of the federal Constitution, which many thought failed to authorize laws restricting individual decisions about what to put in one's own body, but it was also because there were no detectable levels of "recreational" drug use in 1840s America. The lack of any noticeable drug scene was not because no one was getting high; quite the opposite at a time when the newest craze, morphine, had found its way to practically every street corner in the U.S. But no practice of "recreational" drug use was detected in mid-19th century America simply because no drug use was regarded as recreational.[133] Desperate housewives took morphine for "medical" reasons (and often at the insistence of their husbands, who wanted more docile companions), and even alcoholics often claimed their drinking habit had medical value.

Fitz Hugh Ludlow, by contrast, set out on a quest of self-exploration, determined to discover the true nature of the human spirit - which in the Christian worldview that dominated American life, was nearly heresy. According to his biographer, "Fitz Hugh seemed to need [from drugs the] ability to allow him to commune with the inner spirit that energized the material world." Although fully conscious that his quest was blasphemous in the eyes of his domineering father and of society at large, the young Fitz Hugh set out to find the mysteries of spirit not in a book nor in a sermon, but within himself. And in pioneering fashion, he used drugs as a means of self-discovery.

Yet the intrigue of one drug after another ultimately yielded to disappointment, as each eventually showed its limitations. After almost a year of daily trips to the apothecary, (apparently undetected by his parents) the young man stopped by Anderson's eager to try something new, but fearing he had exhausted all available options. But this time he was in luck: his friend and dealer had just gotten a shipment of an exotic new drug from Asia: "extract of cannabis indica," described as "a preparation of the East Indian hemp, a powerful agent in the case of lockjaw."[134] The inquisitive teen found further information in a copy of Johnston's Chemistry of Common Life which Anderson kept around the shop. There Ludlow found a canny description: "Our common European hemp (Cannabis sativa) so extensively cultivated for its fibre, is the same plant with the Indian hemp (Cannabis indica) which from the remotest times has been celebrated among Eastern nations for its narcotic virtues." Describing the drug's effects, Johnston related that while small doses give "great mental cheerfulness," users of larger doses "find themselves almost transported to the scene of the numberless marvels which the Prophet has collected in his paradise."[135]

This description, which riveted the young psychonaut, combined the scant scientific knowledge then available with ancient Eastern myth. Accounts of cannabis' use medical use can be found in the Vedas, some of the oldest surviving works of literature. The oldest surviving treatise on Chinese medicine lists cannabis and describes many diverse applications for it. Arab and many African cultures have employed it as a medicine for well over a thousand years.[136] While no one can say for sure when the first witch doctor or shaman prescribed cannabis, the evidence suggests that human use of the plant as medicine dates back to prehistory. Americans adopted it early on as a home remedy – Thomas Jefferson used it to relieve his migraine headaches – and research in the German universities indicated the use of cannabis, and particularly its seeds, for a variety of indications. Even the English discovered early that *bhang*, a hasheesh extract from India, could relieve many common symptoms, as illustrated by the example of Samuel Taylor Coleridge.[137] But the moment of cannabis's most triumphant arrival in Western medicine came when a study by an Irish doctor named William B. O'Shaughnessy took the Academy by storm.[138] In his exhaustively-titled "On the Preparations of the Indian Hemp, or Gunjah (Cannabis Indica): Their Effects on the Animal System in Health, and Their Utility in the Treatment of Tetanus and Other Convulsive Diseases," Dr. O'Shaughnessy acknowledges from the beginning the apparent paradox of cannabis to temperance-minded Europeans:

> The narcotic effects of Hemp are popularly known in the south of Africa, South America, Turkey, Egypt, Asia Minor, India, and the adjacent territories of the Malays, Burmese, and Siamese. In all these countries Hemp is used in various forms, by the dissipated and depraved, as the ready agent of a pleasing intoxication, [but] in the popular medicine of these nations, we find it extensively employed for a multitude of affections.[139]

O'Shaughnessy published his paper just one year before the commencement of the Opium Wars, when Chinese dependence on the drug was nearing its peak. Newspapers in London and New York made the most of the sensationalist and alarmist potential of opium addiction, which was always presented in racist terms, implying (if not openly stating) that the wily Chinaman deliberately used his opium pipe as a tool to corrupt and ravish pure white women.[140] But while the products of the gristmills of yellow journalism were exaggerated and bigoted, some of O'Shaughnessy's colleagues had conducted scientific studies of opium use, reporting that opium abuse did indeed lead to addiction and serious health and behavioral problems, even while proper use had definite medical benefits. Surveying cannabis use in Calcutta, where he taught at a local medical college, Dr. O'Shaughnessy recognized its therapeutic potential right away, but he also noted that Indian culture was quite permissive toward its use as a social drug, so he proceeded with caution.

Though his initial approach to the drug was ambivalent, his conclusion about it was enthusiastic:

> The preceeding cases constitute an abstract of my experience on this subject, and which has led me to the belief that in Hemp the profession has gained an anti-convulsive remedy of the greatest value.

The "preceeding cases" O'Shaugnessy refers to include stringent case studies of the application of cannabis to a diverse set of ailments, finding it highly effective in the treatment of rheumatism, rabies, cholera, tetanus, cramps, and delirium tremens – some of which could not otherwise be treated by Western medicine at the time.

The paper caused a sensation.[141] Doctors throughout the western world began finding other therapeutic uses for cannabis, and some of the most exciting research was conducted in the U.S. By 1850 cannabis had been listed in the United States *Pharmacopeia* – the most exhaustive and authoritative doctor's reference on pharmaceuticals available at the time – indicating its effectiveness for anthrax, cholera, convulsions, dysentery, gout, incontinence, leprosy, menstrual cramps, a wide range of mental illnesses,[142] neuralgia, snake bite, tetanus, typhus, tonsillitis, uterine hemorrhaging, and – not least of all – treatment for addiction to alcohol, chloral hydrate, and opium.[143] Peter Squire of London manufactured a popular alcohol extract of cannabis which sold in America as "Tilden's Extract"[144], and tincture of cannabis remained the most popular form of administration in the U.S. for decades to come.[145]

This was exactly the same extract which Fitz Hugh Ludlow found in Anderson's apothecary,[146] and after a brief flirtation he decided to try it. Sneaking ⅔ of a gram from the bottle when Anderson wasn't looking, the brave veteran swallowed the dose "without a tremor as to the danger of the result." Going out to enjoy the ride, Fitz Hugh was disappointed to experience no effects.[147] After stealing three more doses of increasing size and feeling nothing at all, Ludlow had just about concluded that he was immune to its effects but tried it one last time to be sure, taking a massive 2 gram dose right after tea.[148]

Ludlow went right away to visit a close friend in Poughkeepsie, enjoying conversation and music for several hours but still feeling no change in his mental state. Disappointed, Fitz Hugh decided to give up the experiment for good when suddenly:

> Ha! what means this sudden thrill? A shock, as of some unimagined vital force, shoots without warning through my entire frame, leaping to my fingers' ends, piercing my brain, startling me till I almost sprang from my chair... I could not doubt it. I was in the power of the hasheesh influence. My first emotion was one of uncontrollable terror -- a sense of getting something which I had not bargained for. That moment I would have given all I had or hoped to have to be as I was three hours before.[149]

Indeed, Fitz Hugh had taken so powerful dose that soon enough he felt that a part of his spirit left his body to observe his intoxicated state, and that the formerly inconsequential distance between his chair and the table had become "an infinity of space." Excusing himself from his friend, Fitz Hugh set out into the dark Poughkeepsie streets. It proved to be an epic journey:

> I dwelt in a marvelous inner world. I existed by various turns in different places and various states of being. Now I swept my gondola through the moon-lit lagoons of Venice. Now Alp on Alp towered above my view, and the glory of the coming sun flashed purple light upon the topmost icy pinnacle. Now in the primeval silence of some unexplored tropical forest I spread my feathery leaves, a giant fen, and swayed and nodded in the spice gales over a river whose waters at once set up clouds of music and perfume. My soul changed into a vegetable essence, thrilled with a strange and unimagined ecstasy.[150]

But before he had gotten home, Fitz Hugh's, (and probably America's) first deliberate hash trip had gone bad. Fitz Hugh encountered a hallucination of a man with a face of "ferocious wickedness," who grimly demanded of him, "You shall bear my burden for me," and tried to hold him captive. Fleeing at a full run, the terrified teen sought out a doctor, who declared there was nothing wrong with him and sent him home with sleeping powder.[151] Home he went, and he made off to bed, but, "[t]he moment I closed my eyes a vision of celestial glory burst upon me." The youth's vivid imagination floated from one paradise to another, taking in unusual sights of every kind but "[t]hrough whatever region or circumstances I passed, one characteristic of the vision remained unchanged: peace - everywhere godlike peace, the sum of all conceivable desires satisfied."[152]

Fitz Hugh had finally found what he had so desperately sought - his soul had been deeply moved, and a revelation of peace had come to him, over him, through him.

He woke to "actual morning and not some hasheesh hallucination,"[153] carefully testing for any lingering bad effects and finding none. Claiming as always that the trial had been a mere experiment and nothing more, he declared the evening a success and swore off any further doses. Yet:

> Had the first experiment been followed with depression, I had probably never repeated it. For days I was even unusually strong; all the forces of life were in a state of pleasurable activity, but the memory of the wondrous glories which I had beheld wooed me continually like an irresistible sorceress.[154]

Despite his intentions, Fitz Hugh eventually found himself back at Mr. Anderson's, where he cultivated a growing drug habit. Drawn by compulsion, Ludlow bravely set out where none had gone before, plumbing the depths of the psyche with his chemistry set in hand.

Apparently his studies didn't suffer,[155] as he enrolled at Princeton in the fall of 1854, where he continued to take hashish and earn respectable grades until his dormitory burned down.[156] The mishap proved fortuitous, as it provided Ludlow with the opportunity to transfer to Union College in Schenectady, New York, an institution which then enjoyed a reputation equal to that of Harvard or Yale.[157] There, his father hoped, Fitz Hugh would develop a close relationship with President Eliphalet Nott, a revered teacher with a reputation for setting wayward youth back on the straight and narrow path.[158] Nott and Ludlow quickly developed an academic friendship and held many discussions on spirit and religion deep into the night. Nevertheless, Ludlow's primary teacher was still hashish. He obtained a new source of cannabis extract in Schenectady and resumed its use, continually raising the dose and frequency until his life became "one unbroken yet checkered dream" - the late-night philosophical discussions with Professor Nott began to take on a new quality. After joining the Kappa Alpha Society,[159] he even inducted several of his brothers into the drug's mysteries,[160] becoming the first observer in Western society to note that subtle changes in the user's mindset give rise to profound changes in their subjective experience under the influence of the drug, and from this insight he pioneered the first techniques to "talk down" a user who has taken too much.[161]

But while Ludlow's attempts to approach the ultimate truths in his visions grew more desperate, his latent guilt rose ever closer to the surface. By Ludlow's senior year, God himself appeared regularly in his hallucinations, but in a form "always menacing, wrathful, or avenging."[162] After this wrathful visage grew to superlative intensity through successive trips, Ludlow finally had to face the most troubling vision of all - a cold emptiness, occupied only by a disembodied voice declaring that God had died.[163] Faced with these horrors and paradoxes, the wayward Christian began to have Messianistic delusions:

> I was invested in a grand mission to humanity, and slowly it dawned upon me that I was the Christ, come in the power and radiance of his millennial descent, and bearing to the world the restoration of perfect peace. I spoke, and it was done: with a single sentence I regenerated the Creation.

When the confused teen fell back to mundane reality from his place of absolute power in the cosmos of his visions, depression followed and suicidal thoughts crept in. His sensitive soul had begun to falter under the weight of his burdens, alongside a fertile imagination amplified and darkened by a titanic drug habit. Ludlow's fate then was even worse than it would be today, for then there was no one else around with whom the intrepid explorer of the depths of the psyche could empathize. Fitz Hugh Ludlow had become lost in the profound solitude of the trailblazer.[164]

During the summer after his graduation from Union, while visiting his family in Poughkeepsie, that bubble of solipsism abruptly burst. Picking up a copy of Putnam's Monthly Magazine at a local bookseller's shop, young Ludlow read the travelogue of an anonymous author who had lived for years among the nomadic tribes around Damascus.[165] During that time, the writer had sampled hashish several times, according to the custom of the locals. He described fantastical visions in exquisite detail, closely matching Ludlow's own subjective experiences, indulging the imaginative fascination of the reader even while ultimately warning against the author's vice, giving the piece a confessional feel in the tradition of DeQuincey's phenomenal *Confessions*.[166]

The masterfully written book had spawned at least a dozen imitators, of which none attained either DeQuincey's literary accomplishment nor his commercial success. Among the many attempts, the Putnam's piece stood out for its capable writing, exotic locale, and faithful depiction of the drug's effects.[167] Reading the author's descriptions of the hash trip, Ludlow could have no doubt but that there was someone who understood, and when he read the anonymous writer's claim that he had successfully sworn off hashish forever, Fitz Hugh's heart lifted in hope.

Ludlow wrote to Putnam's and discovered that the mystery author was one Bayard Taylor, a popular travel writer based out of the States.[168] Ludlow wrote to Taylor, begging for guidance in breaking free of the habit which held him fast in its grip. Taylor, recognizing a kindred spirit and a literary talent, encouraged Ludlow to write whenever he felt the drug's pull.[169] Ludlow took the advice, penning his own magazine piece "The Apocalypse of Hasheesh" which Putnam's published in December of 1856.[170] Ludlow's piece built in the memes present in DeQuincey and Taylor, titillating the reader and indulging her vicarious curiosity while ending on a moralizing note of caution and claiming forthrightly that the author had, like Taylor, given hash up for good.

In Ludlow's case, it was a lie.[171] His habit kept calling so intently that Fitz Hugh, following Bayard's advice, turned out reams of writing at a prodigious pace, expanding the article into a book-length manuscript in only four months.[172] By summer of 1857, Putnam's had received so many enthusiastic responses that they decided to print the work, which they published without a single revision.[173] The book, called The Hasheesh Eater, Being Passages from the Life of a Pythagorean (1857),[174] was a hit, going to a fourth printing before the Civil War put a sudden halt to production.[175] Its author became a literary star, and the American drug counterculture was born.

All of this was great for Ludlow, who turned a decent profit from the book,[176] but the effect on the American consciousness was more profound - as countless Americans became acquainted for the first time with the notion of using pharmaceutical drugs for fun. [177] John Hay, who would become Abraham Lincoln's personal assistant and Theodore Roosevelt's Secretary of State, experimented with the drug after reading Ludlow's book at Brown University.[178] The experiment apparently ended happily for Hay, but not all impressions were pleasant; Ludlow's feverish descriptions of hellscapes all too aptly described the experiences of many Americans who had taken an accidental overdose.

The book reviews of the day invariably compared Ludlow to DeQuincey and, inevitably, hashish to opium.[179] Worse, when describing this mostly unfamiliar drug to a curious audience, many reviewers also mistakenly described hashish as a narcotic, implying that hashish belonged in the same class of drugs as opium.[180] The guilt-ridden imagination of a brilliant, rebellious mind had teamed up with a class of ignorant but prominent reviewers to seed the American consciousness with the idea of congruity between two very different drugs.

Nowhere did these differences become more pronounced than in the laboratory. Morphine, isolated as early as 1805, easily dissolved in saline solution and delivered in precise doses through a fast-acting syringe; but the complex chemistry of cannabis resin eluded doctors well into the twentieth century, allowing only for alcohol-based extracts of inconsistent potency absorbed through the long metabolism of the stomach. This led to countless problems. Whereas patients like Ludlow had to take their best guess at dosage when swallowing Tilden's extract, risking paranoia-inducing overdose as well as ineffectual underdose,[181] patients who received morphine injections experienced a more or less immediate effect of consistent potency, since every injection from a bottle of carefully titrated morphine equaled every other in strength. So it was not morphine's addictive power alone that let it displace hashish as a drug of choice for middle class America in the last decades of the 19th century.[182]

Even so, a little addictiveness can go a long way. Quickly realizing the potential for profit in mass addiction to drugs, a number of home-grown mail-order pharmaceutical companies sprang up, usually founded on a dearth of actual medical knowledge and a surfeit of greed. Packaged in reassuring boxes painted with pastoral domestic scenes, these "patent medicines" went by names like "Mrs. Winslow's Soothing Syrup,"[183] revealed no hint of their active ingredients, and promised efficacious applications as diverse as cough suppression and sleep aid for fussy infants. An already-urgent drug epidemic became even worse, as many Americans (as many as 1% of the population, by some estimates)[184] became addicted to morphine without even realizing it.

One of the first to sound the alarm was none other than Fitz Hugh Ludlow. His plaintively-titled article, "What Shall They Do To Be Saved?" in the August 1867 issue of *Harper's Monthly*, was the first article to be published in a major American journal which warned of the growing opiate epidemic.[185] Its timing was perfect; only two years after the end of the Civil War, opiate use had exploded among war veterans who used morphine as a panacea for lingering traumas both mental and physical, leading to withdrawal symptoms being so common among them that morphine addiction would become known as "the army disease."[186] Ludlow's article exposed the addictive properties of morphine and the vast international distribution network (including the East India Company, German pharmaceutical firms, and dozens of smaller players) which grew out of the Opium Wars into a mature and sophisticated cartel. Ludlow spoke from sad experience, having tried opium first at Mr. Anderson's apothecary during his youthful experimentation and having probably continued its use later in life as his tuberculosis worsened. Given his chronic struggle to give up the habit of hashish (which is less addictive than morphine or even natural tar opium), he must have been no stranger to the seductive but strident call of the opiates. Describing the case of an anonymous "friend" who had begun taking opium to treat a chronic condition, Ludlow cautioned, "There are certain men to whom opium is as fire to tow, and my friend was one of these... The physical power of the drug over him he only realized when attempting its abandonment."[187] Ludlow described the horrible symptoms of withdrawal and concludes on a pessimistic note: "God seems to help a man in getting out of every difficulty but opium. There you have to claw your way out over red-hot coals on your hands and knees, and drag yourself by main strength through the burning dungeon-bars."[188]

The article broke ground in more ways than one. Besides blowing an early whistle on the dangers of the growing drug epidemic, it also represented the first popular call for a paradigm shift in the public perception of the opiate addict, declaring that "such a man is a proper subject, not for reproof, but for medical treatment."[189] He even advocated for a special inpatient facility to be set up only to treat addicts, an idea almost seventy years before its time. To the morality-based Temperance movement of his father's day, still steadily growing in popularity in 1867, Ludlow's ideas were tantamount to heresy.

Still, he wasn't done. Ludlow went on to describe the available treatments, a truly depressing list. Whereas chloroform and ether, two of the most popular drugs of the day, "induce death in nine cases out of ten," other common drugs like nitrous oxide simply failed to help. Eventually despairing that "[t]here is nothing in the faintest degree resembling a substitute for opium" for the withdrawal sufferer, Ludlow nevertheless coyly hinted that "various alleviatives, which can not be discussed in an untechnical article, may be administered with benefit."[190]

The specifics of Ludlow's prescription would not be revealed until the following year, when he edited a book titled The Opium Habit, which compiled excerpts from Coleridge and DeQuincey, a reprint of "What Shall They Do To Be Saved?," and a new chapter by Ludlow, "Outlines of the Opium-Cure." In that chapter, Ludlow got very specific indeed, recommending "cannabis indica extract... as prepared by Hance & Griffith" as a palliative for the awful symptoms of withdrawal ("the earliest effect," he wrote, will be a cerebral stimulus, sufficient to divert the mind from the body's sufferings during day-light").[191] Sensibly, he recognized that hashish could only provide part of the cure, the rest supplied by the removal of the patient from their usual surroundings and placement at a treatment resort, where they could engage in daily exercise and learn new skills (correctly recognizing that part of the grip of addiction is caused by associating with the same people and circumstances which led to use of the drug in the first place).[192]

The book set off a firestorm of controversy. The *New York Times*, reviewing it in October, declared that "men will always be in the habit of using stimulants of one sort or other."[193] The *Atlantic Monthly*, not to be outdone in the pessimism department, maintained that "[i]nveterate opium-eaters generally cannot be cured; their attempts at reformation end in death."[194] But Ludlow refused to give up, promoting his holistic cure even while searching for a better one. The pharmacological crusader knew his plan's limits. The average opium-addicted American didn't have the luxury of time to take off work and family duties to attend a weeks-long retreat far removed from her daily environs, where a careful hashish administration could wean her slowly off her long-seated habit. Until his death, Fitz Hugh continued to search for a permanent cure.[195]

Ludlow was not alone in his search. As the scope of the problem became clear to the American public, a host of compassionate doctors went to work on finding addiction cures. Of all the members of this pantheon of saints, none worked more passionately nor attained greater notoriety than an insurance salesman from Georgia named Charles B. Towns. Towns, having already sold "more life insurance than any other man south of Mason and Dixon's line," moved to New York City to try his hand at investment banking. After the brokerage firm he worked for went under, the disappointed Towns looked about for another venture. One found him when a furtive figure approached him and whispered, "I have got a cure for the drug habit, morphine, opium... any of 'em. We can make a lot of money out of it."[196] The entrepreneurial Towns decided to give it a try. After placing ads in local papers, Towns and his associates forcibly restrained those who responded in a Manhattan hotel room, where he loaded them up with hyoscyamus, belladonna, castor oil, strychnine, and massive quantities of morphine until he finally found the telltale sign of the patient's cure: a copious, green, bilious diarrhea. To Towns' everlasting joy, 90% of all patients subjected to this experience never returned for a second treatment.[197] Charles B. Towns triumphantly announced his new cure, proven 90% effective.[198]

Somehow this baldfaced quack attained a measure of authority in the narcotics debate and a respectable fortune from his regimen of "diarrhea, delirium, and damnation," crowding out most other cures. As for the Fitz Hugh Ludlow theory of the cannabis palliative, he roundly denounced it, declaring "there is no drug in the Pharmacopeia today that would produce the pleasurable sensations you would get from cannabis" and arguing for laws restricting its use.[199]

Other addiction cures soon arrived on the scene. Dr. Edward Huse of Rockford, Illinois described "A New Cure for the Opium Habit" in an 1880 issue of The Therapeutic Gazette: liquid extract of cocaine, as provided by Parke-Davis.[200] In Europe, a young Viennese doctor named Sigmund Freud read the early research with great interest.[201]

At that time, Freud had yet to make a name for himself; at the age of 28, he had already been betrothed to his sweetheart Martha Bernays but could not yet afford to marry her. After trying a sample of cocaine himself,[202] Dr. Freud was deeply intrigued at the opportunity to introduce a new drug to the world medical community, one which he believed could replace opiates for many applications. Giving it to his morphine-addicted friend Ernst von Fleischl-Marxow, he observed that the addict embraced cocaine "like a dying man" and soon showed no more interest in morphine.

Freud knew he had a hit on his hands.[203] In his loving 1884 panegyric Uber Coca he described his experiments with patients like Fleischl-Marxow which led to them escaping the bonds of their morphinism.[204] This did much to promote the early use of cocaine in America and Europe. Two years later, the imminent Dr. William Alexander Hammond recommended cocaine in glowing terms to his peers, confidently declaring it an effective addiction cure while assuring his audience that its habit-forming properties were no worse than those of caffeine.[205] But Freud himself quickly grew to regret his promotion of cocaine as he watched his friend Ernst Fleischl-Marxow trade one addiction for another, kicking morphine for good and carrying a cocaine habit with him all the way to an early grave, seven years after seeking his trusted friend's medical assistance.

In such a whirlwind of claimed solutions to the opiate problem, the Ludlow cure ran no small risk of fading into obscurity; yet some well-known physicians took notice. Dr. Brown's influential An Opium Cure was in no small part inspired by Ludlow's prescriptions, mimicking The Opium Habit's recommendation of using hashish extract as part of a holistic course of treatment. In 1889, E.A. Birch followed up this research with his own article published in *The Lancet* (one of the most prestigious medical journals in the world, then as now) reporting the efficacy of using cannabis to treat not only opium withdrawal but also that of chloral hydrate, another addictive drug much more in use in the 19th century than today.[206] J.B. Mattison confirmed Birch's findings two years later, reporting several cases of morphine addicts who were successfully weaned off opiates with the help of hashish.[207]

But despite the favorable press, hashish extract still struggled to compete in an increasingly crowded field, due in large part to Western medicine's abject failure to comprehend its complex chemistry.[208] A string of successful breakthroughs in isolating alkaloids like morphine, nicotine, and caffeine led the majority of the medical community to expect similar results from cannabis - yet cannabis resin does not contain any alkaloids.[209] Similarly, scientists searching for a single component which could account for the bulk of cannabis's effects - like cocaine or caffeine - would be mystified by the tortuous complications of hashish's chemistry, now known to contain at least 66 cannabinoids and over 400 other constituents with pharmacological actions still not fully understood.[210] Compounding these issues was the imperfect nature of the alcohol-based tinctures commonly available at the time, in which the active cannabinoid contents would settle to the bottom of the bottle, meaning that patients often struggled to gauge the strength of a particular batch, and its composition could vary from dose to dose, even when it came from the same bottle.

A doctor no less prominent than Sir John Russell Reynolds, personal physician to Queen Victoria,[211] discovered a solution to these problems which he published in the Lancet near the turn of the century, but it was too late. While Dr. Reynolds's method was ingenious, it was also complicated and time-consuming, such that most physicians didn't bother.[212] Around the same time, Walter Ernest Dixon presented an audacious proposal to solve the titration problem, pointing out that patients who inhaled cannabis smoke felt the effects of the drug within mere seconds instead of over an hour, making it possible for the patient to titrate the dose himself, relying on the strength of onset to gauge the correct dose. While many doctors spoke with alarm against the notion of a patient inhaling hot smoke - in the name of medicine! - Dixon countered that, in his experience, "hemp taken as an inhalation may be placed in the same category as coffee, tea, and kola [an East African nut containing caffeine]. It is not dangerous and its effects are never alarming, and I have come to regard it in this form as a useful and refreshing stimulant and food accessory, and one whose use does not lead to a habit which grows upon its votary."[213]

Sadly, the movement to use cannabis as a treatment for morphine addiction fizzled, overwhelmed by competition from Charles B. Towns and discredited by an episode in which its originator, Fitz Hugh Ludlow, got taken in by an unscrupulous scam artist. Ludlow, whose health steadily declined in the years after the Civil War, never gave up his dream of finding a final cure for addiction. In the waning months of 1869 he found a promising lead. Learning through a friend that one Dr. Samuel B. Collins of Laporte, Indiana had offered a "miracle cure" which purported to once and finally end opiate dependence, Ludlow wrote the doctor with conscientious haste, noting that he had,

> perhaps, a larger circle of acquaintance with Opium Eaters than anyone else in the country, and have been so happy as to cure a considerable number of the worst cases on record.
>
> None of the cases have, however, I frankly acknowledge, been effected without severe and long protracted suffering... But I have all my life been seeking in vain for some remedy which would act as a substitute and bring the patient out painlessly. Last spring I was almost ready to give up the search in despair - when two of my large circle of Opium correspondents wrote me within a few weeks of each other that you had succeeded in making

> the discovery... If it does all that I understand to be claimed for it, and is itself no form of extract from the accursed poppy - then you have a right to the thankful praise... of every man who loves his race.[214]

Ludlow then asked Collins to express him enough supply to effect one cure, so that he could test it on one of his addicted acquaintances.

After much wheedling, Fitz Hugh Ludlow finally obtained a sample of Dr. Collins' proclaimed opiate cure, which he administered to a friend who had long suffered in opium's hellish grip, noting with his own eyes the immediate improvement the medicine worked. But he couldn't stay to keep an eye on his patient's progress on account of his own rapidly deteriorating health; by 1870 it had become clear that the author had contracted a severe case of tuberculosis, and so he made plans to visit Switzerland, whose doctors were considered at the time to be at the forefront of tuberculosis treatment.[215]

Leaving his friend Henry Read in charge of the administration of the cure which Ludlow had finally become convinced would finally end morphine addiction once and for all, the sickly author traveled to Europe, where he hoped rest would restore him to health and allow him to return to witness in person the impending fulfillment of his life's work. He never got the satisfaction. On September 12th, 1870 in a cabin by a lake in the shadow of Mount Blanc, Fitz Hugh Ludlow, whom both Mark Twain and Charles Dickens called the best of the American writers, died. The official cause of death was tuberculosis, though Harper's Bazar blamed it on a life of hard drug use, including opium, alcohol, and cannabis.[216]

Only after Ludlow's death was it revealed that his great hope, the Dr. Collins cure, was a sham. A chemist friend who examined some of Collins's pills found that they actually contained large doses of morphine, disguised in its effects by added quinine and glycerine.[217] As a result, Ludlow's formerly authoritative reputation diminished, with too many doctors becoming convinced that he had been just another snake oil salesman, and so his holistic hashish cure faded in time from ignominy into obscurity. Although a few pharmaceutical companies did market morphine cures containing cannabis, most of them mixed it in only as an adjunct to the primary ingredient, which was simply more morphine.

And so a great opportunity was missed, and cannabis's promise to break the hold of America's worst addiction epidemic faded into obscurity as stymied researchers went back to the drawing board.[218] Finally, in 1898, the Bayer Chemical Company declared that they had found a solution. By mixing morphine with acetate, they synthesized a new drug which, they proudly proclaimed, would not only serve as an excellent substitute for morphine, but would also cure morphine addiction permanently, finally fulfilling Ludlow's dying dream. Immediately hailed as a wonder drug, the new compound diacetylmorphine became Bayer's flagship pill, distributed without prescription through the mail and at every drugstore, under the trademark 'Heroin'.[219]

And the scientific community cheered, convinced that they had solved the country's drug problem, once and for all.

- 4 -

THE GATHERING OF THE CLOUDS

"We are the drug habit nation."

- Harvey Washington Wiley[220]

"All Mexicans are crazy, and this stuff is what makes them crazy."

-From the transcript of a debate on cannabis in the Texas state legislature, circa 1929[221]

The boy was only twelve when he heard them: the shrieks of agony which would scar his psyche and disturb the course of his career. It happened in 1904, when young Harry Anslinger, Jr., of Altoona Pennsylvania, accompanied his father on a visit with a family friend. A young gentleman of some status in rapidly modernizing Altoona greeted Harry's father warmly in the parlor, polite but strangely preoccupied. Young Harry noticed the absence of the young man's wife who was supposed to join the group for lunch, but noticing that Anslinger Sr. made no comment, Harry kept his mouth shut. Everything seemed to be going well until Harry heard the screams.

The boy had never heard anything so horrific in all his life. A woman's rending sobs of desperate, terrified pain echoed from upstairs, the staccato rhythm jarring the friends out of the smooth habits of their high-society lunch. Her distraught husband excused himself to run upstairs to his young wife's side. Confused and frightened, Harry looked to his father for comfort, but all he saw was a grimace of empathy and sad understanding.

Harry was left waiting and wondering as affairs unfolded upstairs, until finally the stricken woman's husband rushed back down and pressed a slip of paper in the boy's palm, urgently asking Harry to rush it to the pharmacist downtown.

Convinced that the poor woman would die if he failed in his task, Harry hitched up the carriage and spurred the horse through the rough streets. He passed through a town caught up in rapid transition, growing from a quaint frontier hamlet into the home of "new immigrants, rolling farmlands and new factories, miners and roadworkers, foremen and factory leads." The growth of new industry in the agrarian county of dirt roads and farmsteads was noticeable even in Harry's short life, as each year more railroads carved up the rolling green landscape. Harry's frantic carriage ride took him through roads lit by newly installed street lights, burning all night and fueled by an exciting new discovery only a hundred miles to Altoona's northwest: the Drake well outside of Titusville, the world's first successful attempt to drill for petroleum and the beginning of a new age in the world. Harry had even heard that rich folk in bigger cities didn't even need horses, because this new kind of fuel provided enough energy for "horseless carriages" to move under their own power. Even this early, a number of enterprising garage mechanics had begun to bet on this newfangled "automotive" technology, convinced that it was the next big thing.

But none of this was on Harry's mind that day as he frantically searched out a pharmacist before it was too late. Fortunately, access to drugs had never been easier in all of U.S. history; in a totally unregulated market, any American could obtain any drug known to science from outlets as diverse as the local Walgreen's, the J.C. Penney catalog, and the county fair. Harry parked at the closest store, an independent shop downtown, and ran breathlessly inside. He exchanged the paper and cash for a small vial of nondescript pills and rushed his charge back to the concerned adults doting over the sick woman. Her husband thanked Harry curtly but graciously, snatched the pills from his quivering hand, cloistered himself with Harry's father to minister to his wife and left a nervous child to pace the parlor floor downstairs. After a few minutes with the weight of hours, her pain and her screams began to subside, and the tense but relieved pair of men came back down the stairs, gratefully thanking Harry for his quick and competent help.

But Harry Anslinger was inconsolable, struggling already with the terrible significance of what had happened and the role he had played. The bright, literate student saw what was written on the scrip he had carried to the pharmacist, and it didn't take long for him to find out more about what had been in the vial. Twelve year old Harry Anslinger, Jr. had supplied his neighbor with morphine.[222]

By 1904, the patina of ignorance covering the medical community's experience with opium had largely worn off, and doctors at the forefront of medical knowledge had come to a consensus about the dangerous addictive power of opium derivatives, especially morphine and heroin, two of the most popular drugs of the day. Even a twelve year old boy could easily understand how dangerous morphine was, and young Harry was shocked by the blasé complacency of his father's generation towards it. He ruminated endlessly in his quiet Protestant guilt as he relived again and again the episode which made him yet another complicit enabler of a poor addict's suffering. The moral complexities tortured him as much as any physical pain could, and one of the most puzzling questions was how a barely adolescent boy could be sold such a dangerous drug, over the counter, no questions asked. Something was deeply wrong.

September 7th dawned calm and quiet in Vancouver, British Columbia, across the Tacoma Bay from the United States. It was fated to end in wrath. A group of destitute citizens held a neighborhood watch meeting that quickly turned into a protest. As they marched down Hastings Street to City Hall, their numbers swelled to eight thousand. They stood around the steps where speakers railed against the politicians whose negligence had, in their view, allowed the economy to implode. The crowd, now grown to fifteen thousand angry rioters, burned Lieutenant-Governor Dunsmuir in effigy. Encouraged by having reached the critical mass necessary to effect meaningful political reform, they set their sights on the next target. They were ready to restore justice. They were ready to take their country back.

They were ready to kill immigrants.

They caught Chinatown flat-footed. After smashing windows, wrecking stores, and beating down the scattered Chinese who tried to defend their neighborhood, the rioters headed to Japantown. The Japanese, having advance warning, were able to pull together a successful defense and minimize the destruction to their own community. But next morning, in the bloody and smoldering aftermath, many recent immigrants found their livelihoods ruined. The Chinese community petitioned the Canadian government in Ottawa, indignantly noting that their presence had always been welcome when the economy was strong and no one minded that they took the low-paying, back-breaking jobs, such as those in mining and on the railroads, that most whites were unwilling to do; it was only when the job market soured that racist white Canadians made them scapegoats for irresponsible bankers and corrupt politicians. Having pointed out the hypocrisy behind the savagery, they demanded restitution for the crimes against them.

Ottawa sent Deputy Minister of Labor William Lyon MacKenzie King to Vancouver to survey the damage. King, who had ambitions for the Prime Minister's office, faced a dilemma: the Chinese were clearly blameless, and morally the circumstances demanded compensation for them. But if King came to be known as the one who opened the Canadian treasury to pay reparations to Asian immigrants, he would surely alienate the bigot vote, which in 1907 Canada was impossible to ignore. What was King to do to keep his career on track?

Faced with no other choice, King turned to the oldest trick in the politician's playbook: he deflected blame. In his report to Parliament, King downplayed the damage to most of the dozens of Chinese-owned businesses and focused like a laser beam on the opium dens which had been trashed by the mob. Both of them. Noting that government should not in any way support "an industry so inimical to our national welfare," King recommended that Ottawa "render impossible, save in so far as may be necessary for medicinal purposes, the continuance of such an industry" - proposing the distinction between "medical" and "recreational" drugs that has remained at the core of Western drug policy ever since. But he wasn't done. In 1908, King authored another report called "The Need for the Suppression of the Opium Trade in Canada," in which he first began to play on white fears of exotic sexual predators, writing, "The habit of smoking opium was making headway, not only among white men and boys, but also among women and girls." It worked like magic. That same year King wrote, lobbied for, and helped to pass Canada's Anti-Opium Act of 1908, the first national anti-drug legislation in North America, and a model for many more laws to come.[223]

As a political maneuver, it was brilliant. By focusing on the drug use of a minority community, Minister King used an investigation to compensate victims of racial scapegoating as a platform for yet more racial scapegoating; by using the proxy issue of opium to harness the same bigotry that motivated the Vancouver mob without coming off as bigoted himself. It wasn't racism, it was getting tough on crime - but if all the opium dens happened to be owned by the Chinese, so be it.[224] The fact that smoking opium was only a crime because King almost single-handedly made it one was conveniently ignored, and his maneuver took MacKenzie King all the way to the top. He would become Prime Minister in 1921.

At a time when the total population - every woman, man, and child - of the United States totaled less than 84 million, flooding the mail boxes of the U.S. Congress with letters signed by over a million registered voters was no small feat of politicking, yet that's just what Alice Lakey did in 1906. A new movement calling for food and drug safety had been gaining momentum in America, drawing inspiration from the likes of Upton Sinclair's muckraking masterpiece, The Jungle, and by the turn of the century it had coalesced into a political program with definite aims. One of the movement's early all-stars, a gifted chemist named Harvey Washington Wiley, had received a

$5,000 grant (about $140,000 in 2012 dollars) from the U.S. Congress in 1902 to study the issue of food safety and to make recommendations. His report, based on empirical work identifying common food ingredients and testing their effects on human subjects, may have lacked Sinclair's literary flair, but nonetheless his results thundered through the halls of power.[225] Wiley's research revealed a host of harmful chemical adulterants and foul ingredients in the American food supply, present at times in alarming levels; his work set off an inexorable drive toward legislative action. And in one of modern political history's canniest moves, Wiley teamed up with the activist Lakey to drive the point home.

Lakey, born in 1857 in Shanesville, Ohio, lost her mother at the age of six. Her father, a Methodist minister and insurance broker, recognized young Alice's remarkable singing talent and sent her with a private tutor on a European singing tour lasting almost a decade. Upon her return to the States, Lakey, now a young woman, settled with her father in Cranford, New Jersey, where Alice joined the ranks of a new breed of American woman: independent, thoughtful, and above all politically active. The image of the politically active woman was, in fact, only one facet of the new Progressive movement rapidly making its mark on American politics, combining muckraking exposes with demographic shifts to tip the political scale. Lakey came up on the leading edge of the Progressive movement, and within a few years, she had convinced the male politicians of Cranford Village to build a public school, fund a public fire department, and start a clinic for newborn babies.

Early in 1903, when her father fell ill, Alice Lakey found her career's focus. Looking on local store shelves for nourishing foods for her sick father, Alice instead found rotten foods adulterated by strange chemical additives like copper, lead, and borax. Already a savvy activist, this new motivation led her to embrace larger ambitions. She joined the Domestic Science Unit of the Cranford Village Improvement Association, and quickly attained the office of President. Lakey went right to work, writing to Federal Secretary of Agriculture James Wilson to ask him to recommend a speaker on the subject of food safety to address her group and to help devise a strategy to clean up Cranford's food supply.

Secretary Wilson was a busy man, trying to juggle populist food safety outrage with the more typical duties of developing best agricultural practices for the nation at a time when small family farms dominated the American economy. Still, he found the time to arrange for one of his brightest and most passionate chemists to address Lakey's group in Cranford. That chemist was Harvey Washington Wiley.

Wiley delivered a damning report to Lakey's group: the nation was in crisis. All throughout the U.S., Americans obliviously consumed poison-laden food. Wiley's pioneering chemistry work at the Department of

Agriculture had revealed dangerous levels of adulterants on every grocery aisle. Worse, when sick Americans suffering from this diet turned to the medical business for relief, they were given dangerous and addictive drugs like morphine, heroin, or chloral hydrate. These drugs sold under innocuous-sounding names like 'Mrs. Winslow's Soothing Syrup' and bore no indications of their specific contents or warnings about their ill effects.[226] Thus, Wiley warned, an entire generation was slowly being poisoned to death by the reckless pursuit of short-term profits by unscrupulous corporations in an entirely unregulated market for food and drugs.

It was a match made in Progressive heaven. Immediately after the talk, Lakey and Wiley began brainstorming, combining Wiley's encyclopedic technical knowledge of the problem with Lakey's political cunning to build a broader coalition around the issue. By 1904, the pair had moved on from the Cranford Village Improvement Association to set the agenda for the New Jersey Federation of Women's Clubs. That same year, Lakey and Wiley approached the largest women's group in America, the National Consumer's League, with plans to form a special unit to investigate food safety issues, with Alice Lakey at its head. The League agreed.

By 1905, Lakey and Wiley had built sufficient support to meet with President Teddy Roosevelt, a maverick Republican with Progressive sympathies. Roosevelt, who would sign the Meat Inspection Act that year, promised Lakey and Wiley to support their proposed bill if they could organize a large-scale campaign to petition Congress for the bill's passage.

Lakey was born for this. Calling in every connection she had cultivated in the National Consumers League and other Progressive organizations, Lakey organized a drive which resulted in more than a million letters, almost all written by women, delivered to Congress. Congress caved under the pressure, and Lakey and Wiley's legislation became law as the Pure Food and Drug Act of 1906.

Lakey and Wiley's campaign succeeded in large part because of earlier pioneers of women's politics who blazed the trails they followed - and of their forebears, none had so great an impact on American politics as the legendary Carrie Nation, whose zealous vigilante enforcement of alcohol prohibition laws in her home state of Kansas provided inspiration to women throughout the country. Nation, who was known for striding boldly into saloons and taking a hatchet to casks of ale, probably had her greatest impact not through vigilantism, but through her successful framing of drug use as a moral issue, something not seen in America since the time of the early Puritan colonists. Pleading on behalf of the countless silent housewives who lived in fear of their husbands' alcohol-fueled violence, Nation made the case: if beating one's wife is wrong, then imbibing a beverage which made one more likely to beat one's wife must be wrong, too. This logic allowed Nation's coalition, the Carrie Nation Prohibition

Group, to convert millions of American men to their cause while simultaneously sowing the seeds for women's suffrage.

The Temperance Movement also benefited from the xenophobia of the time. As waves of new immigrants arrived at American shores, the Anglo Protestant "dries" of the Temperance movement grew in number, successfully playing on distrust of the mostly "wet" recent immigrant population from Catholic countries, such as Ireland and Italy, whose cultures embraced alcohol.[227] More and more Americans began to see drug use, formerly an entirely medical issue, in moral terms. More significantly, an obsolete idea from the earliest days of American colonial history returned to politics: criminal punishment for behavior undesirable on the basis of moral disapproval alone.

While the notion became applied to a few different drugs (some states even criminalized cigarettes), the most prominent banner rallied around the scourge of alcohol and its tendencies to awaken latent violent tendencies. The first laws of the Temperance Movement mostly passed in Midwest states in the late 19th century: Iowa, North Dakota, and Carrie Nation's home state of Kansas. But in 1907, right at about the same time as Lakey's historic letter drive, a wave of Southern state legislatures buckled under pressure from temperance advocates, criminalizing alcohol in Mississippi, Alabama, Georgia, and the Oklahoma territory. North Carolina and Tennessee would soon follow, in 1908 and 1909.[228]

Yet this raft of anti-drug legislation did little to console Harry Anslinger, still haunted by memories of narcotics and obsessed with their dangers. While he was still in his teens, he saw the horrors of opiates encroach again on his psyche, this time ensnaring a friend of his - "a young pool player, the best in Altoona, a bright-eyed, grinning youth" who "sang tenor like an angel."[229] Yet Harry watched in horror as the youth's beautiful voice became marred by a new vice just beginning to captivate the white strata of American society: smoked opium.

So in fact William Lyon King's fearmongering was not without basis; Harry's friend in Altoona represented one of the earliest cohorts of young whites to seek out the thrill of smoking tar opium, as Chinese immigrants took it in their dens.[230] Most historians agree that the primary driver of this shift was not the inscrutable influence of Chinese immigrants, as King implied, but rather an ironic consequence of laws ostensibly passed to protect America's youth. Beginning in the late nineteenth and early twentieth centuries, a new rebellious generation, popularly known as the sporting class, built a lifestyle around seeking out taboo experiences, especially illegal ones. At the sporting class's core were marginalized bohemians, such as touring theater companies and circus performers, but by the turn of the century the sporting class expanded to embrace certain upper-class youth who were attracted by opportunities to rebel against the values of their parents. Thus, San Francisco's innovative anti-opium den ordinance in 1875, passed to harass Chinese immigrants, actually caused opium use to rise among white youth.[231] In 1892 one of the city supervisors lamented that "the vice is spreading amongst depraved white people of both sexes… carried on in private houses or in rooms secretly kept by white people."

By 1907 the habit had spread so much among the youth that white Americans (typically aficionados of morphine, which is far more dangerous than tar opium) clamored for even more punitive legislation against the Chinese. Congress had already passed the Chinese Exclusion Act of 1882, intended to prevent immigrants from taking jobs from "hardworking white men". But as the scandal of opium smoking spread, racist rabble-rousers began to look to more creative solutions to discriminate against their hated enemies.

The freshly-passed Canadian Anti-Opium Act (spurred by the 1907 Vancouver riots) provided an ideal model. Policy wonks began to draft federal laws which, like the Canadian law that inspired them, drew a sharp distinction between prohibited Chinese use of opium and permitted white use of morphine, by labelling the relatively safer opium as a "hedonistic" drug and the far more addictive morphine as "medicine". This distinction allowed the government to persecute the Chinese for smoking in opium dens (which was "recreational use") while giving millions of addicted white Americans a pass, simply because they got their supply from a doctor, drug store, or mail-order catalogue.[232] In light of the massive waves of white opioid addiction at the time, the only real crime the Chinese committed was diverting business away from the lucrative opiate pharmaceutical industry, which was in the process of rapidly consolidating in the early 20th century.

In no small part, this consolidation resulted from the passage of the Pure Food and Drug Act. The Food and Drug Administration which the law created (under the direction of Harvey Washington Wiley, its first administrator) had been so successful at curbing the worst abuses of the U.S. pharmaceutical industry that nearly all of the most unscrupulous snake oil salesmen shuttered their doors overnight.[233] Few were concerned with stricter federal legislation during the first few years in the wake of its passage. Nevertheless stricter legislation came, building on the successes of the FDA to consolidate pharmaceutical control. Strangely, the agency which lobbied hardest for this move was not the FDA, but the State Department.[234]

The Opium Wars had set the ancient, rigid Chinese culture adrift in vast new world of globalization. By 1900 the leaders of the proud empire who had proclaimed that it needed nothing from the "barbarian" races populating the rest of the world had turned to trying to gather up the tatters of its once vibrant economy. They recognized that they needed help: it was time to open China to the world markets. Yet the last time China had done so, its brutal exploitation by the East India Company seared a strong anti-imperialist sentiment into the collective unconscious, so Chinese diplomats could only cautiously signal their willingness to gradually open up trade.

Every imperialist power in the world, from Japan to Great Britain, got giddy over the thought. The Chinese population was so vast that, as one American folk observation went, if every Chinese person exchanged their traditional footwear for American-made shoes, every factory in the country could run at full capacity for years straight and still not fill the order.[235] The Americans, coming late to the imperialist party, pursued Chinese business opportunities with a desperate aggression that stood out among the other delegations, so much so that it touched a tender nerve with Chinese diplomats still smarting with the memory of the Opium Wars. (The fact that the Chinese Exclusion Act of 1882 was still on the books didn't help matters either.) American business interests needed a way to soothe some ruffled feathers.[236]

Charles Henry Brent, an Episcopal Bishop who had traveled to China to oversee the church's missionary work there, believed he found a way to assuage the Chinese concern with a show of American good faith. Noting the enormous rates of addiction to opium in the country (a habit practically forced on the Chinese by the British Empire through the Opium Wars), Bishop Brent found that the Chinese Imperial government was still extremely resentful of opium smoking and considered it one of the worst impediments to China's economic recovery and entry into the modern world. Writing to President Theodore Roosevelt in 1906, Brent described with disgust the epidemic of opium smoking in East Asia, and urged Roosevelt to consider a conference of countries involved in the international narcotics trade, as a way to help China out of its problem.[237]

Roosevelt approved the idea, ordering Secretary of State Elihu Root to execute the plan. The Chinese, eager to end the opium scourge, agreed to host the conference in Shanghai if the Americans could arrange the participation of a sufficient number of countries involved in the opium trade. England, the Netherlands, India, and Persia - four of the world's biggest players in opium - joined America and China; Turkey, a major opium exporter, declined. The Shanghai Commission convened on February 1st of 1909, with Bishop Brent at its head - a diplomatic choice which signalled that America was pursuing a moral course in leading the crusade against opium. Joining Brent on the American delegation was Dr. Hamilton Wright, a dashing Washington, D.C. physician who had had the luck or foresight to marry Elizabeth Washburn, a well-connected D.C. activist and socialite who helped her husband secure the appointment.[238] While both Brent and Wright believed fervently in the necessity of curbing the international trade in narcotics, their methods differed greatly. While the congenial bishop tried to build consensus even if it resulted in compromise, the moralizing doctor bluntly refused to budge an inch.[239]

Wright's steamroller diplomacy would cause trouble for the American delegation several times, but by far the greater hindrance was the legislative situation at home. In the months leading up to the historic conference, the American delegation realized with a rude jolt that the United States, the very country which had proposed the anti-narcotics conference, had no federal anti-narcotics legislation of its own. While the U.S. had no lack of political will, the Constitution did not grant Congress the authority to impose such a law.[240] Brent and Wright foresaw that the other Shanghai delegations might not appreciate the subtleties of the American system of limited government, seeing only hypocrisy. The diplomats frantically cabled the State Department, urging Congress to pass at least token legislation against opium on the federal level.[241]

Back in Washington, Harvey W. Wiley noted the delegates' concerns and advocated for the use of the Commerce Clause, which gave Congress the power to regulate commerce which crossed state or international boundaries - at the least, a ban on opiate imports could be constitutional.[242] But domestic doctors and pharmacists still relied on opiates such as morphine to treat patients, and vigorously opposed any federal interference in their ability to do so. Finding Wiley's proposal to be politically impossible, Congress compromised, passing a ban on only smoked opium while preserving a loophole exempting imports for "medical purposes."[243] Ironically, the predominant people in America to regularly smoke opium were Chinese immigrants - so the law intended to engender goodwill among the Chinese only ended up increasing their persecution.

The Chinese delegation was not impressed; the Commission failed after four months to come to any substantial agreement. Most of the powers which attended threw in the towel and went home, showing no interest in trying again. China still wanted an agreement, but the other stakeholders, even most of the U.S. State Department, had apparently cooled to the idea.

Dr. Wright refused to give up. Despite opposition from his State Department superiors and even from Bishop Brent, Wright jumped up the chain of command, all the way to Secretary of State Philander Knox, who, after listening to Wright's pitch for two minutes gave him the laconic response: "Go ahead."[244] For the next two years Dr. Wright was a whirlwind of energy, imposing his blunt, unilateral 'diplomacy' on the rest of the world powers, eventually getting twelve of the most influential to agree to meet again, this time in The Hague, Netherlands. By the time the Hague convention was scheduled to convene in 1911, Dr. Wright was determined not to repeat the mistakes at Shanghai, and he lobbied early, often, and hard in the U.S. Congress for some kind of federal anti-drug legislation with teeth.

Article VI of the U.S. Constitution states that "This Constitution, and the Laws of the United States which shall be made in Pursuance thereof; and all Treaties made, or which shall be made, under the Authority of the United States, shall be the supreme Law of the Land; and the Judges in every State shall be bound thereby, any Thing in the Constitution or Laws of any state to the Contrary notwithstanding" [emphasis added]. For centuries, courts have recognized in this clause a clear statement of the supremacy of federal law, including foreign treaties, over state law; if ever a conflict arises between the laws of a state and a foreign treaty, the foreign treaty prevails.

In late 1909 Hamilton Wright made the Supremacy clause the linchpin of his plan to circumvent the Constitution and enact prohibitionist domestic drug laws around the following points[245]: first, under Commerce Clause authority, a federal law would require every drug dealer to register with the federal government and to record every transaction in narcotics, under penalty of a heavy fine; second, possession of drugs without the proper paperwork would be presumed a crime; third, and most significantly, state authorities would always have access to the records thus gathered by the federal government. And finally, the teeth: an international treaty to require the states to enforce prohibition, where the federal government alone could not.

In such a treaty's wake, any state which passed laws in conflict with the treaty would run afoul of the Supremacy clause - and thus the federal government could strong-arm the states into doing its own policing. The feds would assist, of course; under Wright's proposed legislation, anyone found in possession of narcotics without the required records would be presumptively guilty under federal law, and anyone who complied with federal law by registering could have their records delivered by the feds to state prosecutors, who would use the defendant's registration as a confession against him. Wright had found a way to hack the system, get around the thorny constitutional issues involved, and create a formally legal regime of federal drug policing and prohibition.

Wright shopped the federal law portion of his plan around Capitol Hill, and on April 30, 1910 Rep. David Foster of Vermont formally introduced it as a bill.[246] The Foster bill was sweeping and punitive, requiring fastidious recording of each and every minute amount of morphine, heroin, cocaine, chloral hydrate, and cannabis dispensed to any patient for any purpose; failure to register or record a transaction resulted in penalties of up to $5,000 (about $125,000 dollars in 2012) and up to five years' imprisonment.[247] A month later Dr. Wright appeared before the House Ways and Means Committee to defend it. There he encountered a sharp rebuke from Southern Democrats, who because of Reconstruction were wary of federal power.[248] But Wright thought he knew their Achilles' heel. "[O]ne of the most unfortunate phases of the habit of smoking opium in this country" was, he sorrowfully reported, "the large number of women who have become involved and were living as common-law wives or cohabiting with Chinese in the Chinatowns of our various cities."[249] Nor were the good white folk living in rural areas safe from the drug menace, because cocaine had recently attained great popularity "among the humbler ranks of the Negro population in the South." While Wright expressed puzzlement as to why more people weren't aware of this dreadful menace[250] (and indeed how the drugs got to remote rural areas in the first place, since "the lower order of working Negro is not willing, as a rule, to go to much trouble or send to any distance for anything"[251]), he urged his audience not to scrutinize this image of their worst fears in light of facts or reason, for "it has been authoritatively stated that cocaine is often the direct incentive to the crime of rape by the Negroes of the South and other sections of the country."[252]

But Wright had misread the zeitgeist. While the doctor tried desperately to pander to the lowest common denominator of the Southern delegation, he failed to appreciate that some of the most compassionate and progressive policies to deal with the national drug problem had been developed in the heart of the South. The most radical changes came in 1912 when the city of Jacksonville, Florida installed a new City Health Officer: a Dr. Charles E. Terry, who had some radical ideas about treating addicts.[253]

After watching previous punitive policies fail to reduce the incidence of addiction, Dr. Terry took bold action, setting up a city-owned and city-run narcotic dispensary where addicts could receive their drugs free of charge and under medical supervision.[254] In so doing, he also created a system whereby the city could collect accurate data on its addicted residents and make recommendations on the best way to help them. Registering about 1% of the city's population, Dr. Terry discovered that many of the popular myths about drug use in the South were untrue. While popular opinion held that the average "dope fiend" was a sexual predator, waiting to ravish unsuspecting women, Dr. Terry discovered that in fact female addicts outnumbered male addicts by a ratio of 3 to 2.[255] And whereas many believed that blacks used more drugs than whites, Terry discovered that twice as many whites sought out the city clinic's services than blacks, even though the white and black population of Jacksonville were roughly equal at the time. New York[256] and Tennessee[257] also experimented with government-run maintenance programs and reported similar results;[258] the case of New York was especially successful, since it reduced the workload of the city's clogged courts, solving another problem by taking addicts out of the criminal justice system and putting them in the care of doctors instead.

This groundbreaking success deep in the South[259] may have swayed the better angels of Congress's nature. Despite Dr. Wright's breathtaking performance, the House was hesitant to pass such an unprecedented expansion of federal power, and the bill remained stalled in committee in 1911, the year that the Hague conventions were scheduled to convene. Thus Wright appeared at his own treaty convention, for the second time in a row, empty-handed.[260] With him went Bishop Brent and an unknown California pharmacist named Henry J. Finger whose political connections to Secretary of State Philander Knox ensured his appointment. The choice mortified Wright, who couldn't imagine why Finger would want to tag along.

The convention opened to lackluster interest from the majority of attendees. The various delegations took turns boring each other until Finger "unexpectedly rose from his seat" to denounce the evils of cannabis, insisting that every nation present criminalize the drug under the harshest possible penalties. Declaring that wholesome residents of his home state of California - especially those who lived in and around San Francisco - had become alarmed by the "large influx of Hindoos... demanding cannabis indica," Finger alleged that these migrants from the Subcontinent had attempted to seduce white Americans with their habit.[261]

The other delegates, who had come expecting a conference on opiates, stared dumbfounded; only Italy and South Africa supported the proposal, and the other nations merely expressed indifference and moved on.[262] Yet despite all the negotiations, the Hague convention closed with no treaty, a failure more than one contemporary blamed on Wright's roughshod diplomatic style. The countries pledged to meet again, which they did, in 1913, again at The Hague, and for a moment Wright appeared close to his triumph; but while the treaty was still two votes short of passage, the assassination of Archduke Ferdinand set off World War I, and the treaty was forgotten. Dr. Wright would never see his dream fulfilled; eventually he disqualified himself from foreign service by refusing to give up an occasional alcohol habit.[263]

Nevertheless, the legislative machinations set in motion back in Washington, D.C. had not lost their momentum, and the determined doctor rushed back home to try to salvage the domestic portion of his plan. Although the Foster bill was eventually killed while still in the House, Wright found another sponsor, Representative Francis Burton Harrison of New York, a Tammany Democrat who never minded a sip of whiskey but hated all narcotics. The Harrison bill, which resembled the Foster bill in almost every respect,[264] nevertheless had better prospects of passage, benefiting from a sweep of the House by a Democratic party comprised of crusading Progressives from the North and Dixiecrats from the South; if Wright could present a pitch combining muckraking aimed at the Progressives and race-baiting aimed at the Dixiecrats, he might shepherd the bill through.[265] Even more crucial was the support of the American Medical Association (AMA) and the American Pharmaceutical Association (APhA)[266], two powerful trade groups[267] which had opposed the onerous books of the Foster bill but would favor a streamlined version of the bill out of a belief that tough federal legislation could wipe out the patent medicine vendors, their mutual competitor.[268]

For the first time, the American medical cannabis industry faced serious legislative threat. So tightly had the fear of any intoxicant been wound around a society obsessed with Temperance that a loving book about cannabis's virtues proved sufficient to spur a raft of restrictive legislation in the northeast.[269] In the spring of 1910 Dr. Victor S. Robinson, a brilliant and passionate physician with an eye, perhaps, to emulate Dr. Freud's successful introduction of cocaine, wrote a hagiography called "An Essay on Hasheesh," which the Medical Review of Reviews, a journal popular in New England, printed in 1912. In his essay Dr. Robinson meticulously summarizes the state of knowledge of cannabis' chemical composition and pharmacological effects (which was very little of either), and the doctor's own experiments with the extract: on dogs, on his friends, and on himself.

Robinson describes cannabis in terms all too like cocaine in Dr. Freud's notorious work, noting its remarkably low toxicity and effectiveness for a host of ailments. Yet the enthusiastic doctor went a bridge too far by including an enthusiastic and vivid description of the hash trip's exquisite pleasures, provoking fear to arise out of the Temperance mindset.[270] Thus, although no cannabis "problem" even existed in the Northeast, Maine, Massachusetts, Vermont, and Rhode Island all passed laws requiring a prescription for "cannabis indica" in the two years following the publication of Robinson's essay.[271] It would appear that mere association between a drug and a state of intoxication was sufficient to ensure its restriction.

The hysteria in New England brought the politics of drugs to the tipping point. After amending the Harrison bill to give the AMA and APhA the simpler paperwork procedure they desired, the House passed it as HR 6282 on June 26th, 1913.[272] The Senate passed its own, somewhat watered down version in August, and various compromise negotiations and procedural matters held up passage of a final version by both houses until December of the following year. Henry J. Finger and other early drug warriors pushed hard to include cannabis in the restricted list, but late in the bill's negotiation opposition from doctors and concerns about constitutionality killed the idea.[273] On December 17th, 1914, with no fanfare (not even a mention in the *New York Times'* daily legislative record), President Woodrow Wilson signed the Harrison Act into law.[274]

Legal experts debated how the Act would be enforced. Some argued that the intent of the bill was merely to collect records and to use them as the basis for convincing state legislatures of the need for action. Others took a dimmer view. Shortly after the passage of the Harrison Act Frank Freericks, a pharmacist and lawyer writing in the Journal of the American Pharmaceutical Association, ominously warned that the newly-adopted legislation would be used by the federal government to encroach on the traditional police power of the states in an attempt to regulate the narcotics traffic.[275] He mocked the naive position of those who believed the "record-keeping" story and joined with those predicting that the provisions for fining non-compliant dealers would be used as a penal cudgel by Washington, D.C. The Journal's own editor, James H. Beal, disagreed, assuring the professional pharmacists of the nation that the only consequence of the Harrison Act would be to collect information.[276] Beal's appears to have been the majority view, with even the Surgeon General of the U.S. opining (in 1915) that nothing in the Act would hinder a physician or pharmacist from continuing to provide morphine and other narcotics dispensed "in good faith."[277]

Yet the disagreement revealed the ambiguity in the Harrison Act's language, which extended not only to questions of enforcement but even to what kinds of treatment would be allowed. For example, the precise meaning of the "good faith" provision of the Act cited by the Surgeon General was not specified, and legal scholars debated whether that provision allowed doctors to pursue the kind of maintenance therapy pioneered by Dr. Terry in Jacksonville. [278]

The Bureau of Internal Revenue (now the Internal Revenue Service) rendered all such discussion moot. Internal Revenue interpreted Harrison as creating *carte blanche* to exercise police power, announcing that it would aggressively punish doctors, pharmacists, and even patients who possessed more than minute quantities of narcotics. The BIR promulgated strict rules for its enforcement of the Harrison Act, clearly stating a zero tolerance position for any maintenance therapy - and began arresting Americans on March 1st, 1915.[279]

The federal government's radical actions were roundly and sternly rejected by the courts, which declared the government's application of the Harrison Act unconstitutional in Pittsburgh, Memphis, Kansas City, and Florida. But the biggest defeat would come in 1916, when the U.S. Supreme Court struck down large portions of Harrison, in particular the Treasury's interpretation of the "good faith" clause as prohibiting maintenance. The Treasury commissioner complained that the decision "makes it practically impossible to control the illicit traffic in narcotic drugs."[280] It seemed as if the principles of small government had won.

But then, only three years later, the Supreme Court dramatically reversed its position. David Musto, one of the foremost authorities on the history of American drug policy, attributes the reversal to a "Red Scare" which occurred immediately after the triumph of the Allies in World War I. The sudden, rapid rise of the Bolsheviks in Russia in the wake of the Great War shocked American capitalists, who feared a similar uprising in the U.S. Pro-capitalist propaganda warned Americans to be constantly vigilant against insidious Communists, and to be sharp and skeptical, if not downright paranoid, towards their fellows so as not to be taken in by one of the Communists' many wily plots for world domination. Needless to say, all activities with a Communist air must be reported to the authorities.[281] The Supreme Court, argues Musto, had merely swung more in line with public opinion[282] when it decided two Harrison Act cases on the same day: the first (*U.S. v. Doremus*) declared that transfers of narcotics were only legal if by "prescription"; the other (*Webb et al v. U.S.*), declared a "prescription" to be legal only if given for something other than the maintenance of an addiction. The vote was five to four.

By the time of the Harrison Act's passage the Temperance movement had attained rare political pull in American politics, poised to pull off the most audacious of democratic coups d'etat, the constitutional amendment.[283] The pending ratification of the 18th Amendment on January 16, 1919, finalized only four years after the Harrison Act, overshadowed all other debates about intoxicants in America, and the Congressional legislation carrying it into effect, the Volstead Act, became the primary instrument for dealing with drugs and the focus of all attempts at reform. Both the Volstead Act and the Supreme Court's blessing of the Harrison Act followed in the turbulent wake of World War I, which made it easier for legislators to urge abstemiousness - forced if necessary - in order to maximize national readiness for war, as if the whole country still belonged under barracks discipline.[284] In any case, the remarkable political achievement represented the legislative high water mark for the Progressive movement's efforts to regulate the substances Americans eat, drink, and smoke. And while the campaign had begun with the best of intentions, (eliminating poisons from the food supply), in Prohibition the government had strayed, in the minds of many, too far into paternalism.[285]

Henry Anslinger could not have been more thrilled. For him, the Volstead Act was a godsend and a windfall, a balm for his faith in the morals of his country and a boon for his pocketbook, too. After the War, the returning OSS hero (who had made off with some personal effects of the Kaiser - the modest spy never said how[286]) settled down with a nice wife with nicer political connections who arranged for her new husband to get a job enforcing the new Volstead Act in the Bahamas.[287] There he tracked down the vernal shoots of the perennial American pastime - smuggling - and loved the work. These were idyllic times, on the surface. The committed drug warrior made a good name for himself in government circles by cultivating useful cooperation from the British colonial governments of outlying islands.[288] His job security seemingly secured, the sensitive soul turned to an old love: writing. He wrote a breezy, reassuring article debunking the common myth of man-eating sharks which got picked up by The Saturday Evening Post - a big break for an amateur author. But a wave of sensational letters from incredulous readers prompted the Post's editorial board to ask Anslinger to revisit the subject in greater depth, and his follow-up piece contradicted every conclusion of the first:

> [A] fifteen-year-old boy died as a result of being attacked by a shark at Port Hacking, Australia... It was found that the flesh of the right leg had been torn completely off from thigh to ankle, leaving the bones exposed and causing death shortly thereafter... [A] shark captured at Koolau, Hawaii, was found to contain human bones and a pair of swimming trunks. The bones consisted of more than half of the upper part of a skull, a hand, a knee, two whole arms, one leg bone and the first and second cervical vertebrae. A quantity of short black hair was attached to the skull...[289]

And so is revealed the young man's ability to pivot 180 degrees to foist the carriage of his career, revealing the nadir of a scarred psyche. And so the cracks begin to show, like a bead of dark blood rising atop the white skin of a pricked forearm from a vein flooded with poison; like a placid, glassy sea roiled by a predator's lunge and darkened by a font of crimson; like a haunting scream of agony ringing in the stairway, piercing a child's veil of security, of knowing there's a paternal figure by one's side - looking out, locking up, and offering calm assurances that everything was going to be just fine.

- 5 -

BERNAGOZZI'S *SCAVEZZATRICE*

"We shall, by and by, want a world of hemp more for our own consumption."

- John Adams, 2nd U.S. President

"All perverts may not be marijuana smokers, but practically all marijuana smokers are perverted."

- Colonel Garland Williams, Federal Bureau of Narcotics

Late in the evening of November 4th, 1928, a bleary-eyed bellhop at Manhattan's upscale Park Central Hotel encountered Arnold Rothstein, America's most notorious gambler cum playboy, clawing his way resolutely toward the exit on Fifty-Sixth Street. Rothstein, trailing a long spreading bloodstain through his trouser leg onto the lobby's carpet, politely requested a taxi. The bellhop summoned medics instead, and with them came the police. Rothstein's position at the top of New York's criminal underground was the worst-kept secret in Manhattan, and the cops, sensing weakness, closed in.[290]

Rothstein held on for forty-eight hours, finally expiring on Election Day, unable to collect the $500,000 he had riding on Hoover. So truly did he hold to the code of the underground that he refused to give up even his own murderer. Perhaps it didn't matter. Ever since Rothstein had risen to the top of America's market for illegal booze and drugs, he had been in everyone's crosshairs.

Henry Anslinger, still stationed in Nassau, noted the death with interest. His family connections and diplomatic skills had taken him far within the Treasury department, and he fully expected that the gangster's death would shake up the upper bureaucratic strata of Washington. As for Rothstein's supposed "gambling winnings," he didn't believe it for a second; everyone knew that New York's most famous playboy was running booze. With the "gambler's" death, the odds favored a muckraker's dream, as either a damning expose of corruption or an extremely public hush-up seemed likely from the city's infamous Tammany Hall; in the end, America got both, as the NYPD detectives managed to "lose" the majority of Rothstein's personal effects and the lid blew open on Rothstein's political inside tracks anyway: the dashing, well-dressed gangster held a vast array of cops, prosecutors, and judges in his pocket, many of whom apparently extended him courtesies not for money but simply to curry his favor. Heads were sure to roll. The tall, handsome Anslinger had demonstrated himself at his rum-running assignment ably, and fully expected to go far within the Treasury's Prohibition unit. Little did he know that quiet developments on the other side of the world would launch his career in an unexpected direction.

The man known as Bernagozzi stopped to wipe his brow. This was unendurable. Slamming an iron hinge over and over again onto stalks as strong and thick as young trees, eventually snapping the woody core; repeating at a dozen points along the stalk, up to thirty feet in length - all that hard work processed only one plant, and the Bernagozzi family farm had acres to brake. The stoic farmer creaked his truculent frame, made old before its time, and ponderously gazed at the barn rafters. There had to be a better way.

That night before bed, he began drawing plans. Bernagozzi had heard of the steam engine, the awesome invention capable of generating enough force to pull long trains weighing many tons; if steam could do that, he reasoned, braking cannabis stalks should be no problem. It was hardly an original idea. Thomas Jefferson filed the first U.S. patent for a device to crack cannabis hurds all the way back in the 18th century; his design used horses for power. Other inventors tried their own designs to take over the backbreaking labor, but none had yet succeeded on a commercial scale.[291]

Bernagozzi's design was a breakthrough. The engine, powered by coal or wood, created the necessary steam pressure to turn a conveyor and claw which worked together to align the shorn cannabis stalks before processing. Once inside the chassis, the stalks encountered a spoked wheel with bricks attached at the end of each spoke. Precisely spaced, each brick crushed the cannabis stalks just enough to snap the woody hurds but not enough to damage the valuable interior bast fibers. A triumphant Bernagozzi dressed these fibers and brought them to market, where the machine-braked fibers were generally judged equal in quality to those processed by hand. The invention, which Bernagozzi called the "scavezzatrice," spread rapidly through Italy and then Europe. Though crude, it still set off yet another revolution in hemp history, creating the possibility that cannabis fibers could compete with cheaper competitors for the first time since the invention of the cotton gin.

By 1900, American cannabis fibers had nearly been priced entirely out of the market. Russian hemp, a major competitor in the early 1800s, gave way to even cheaper fibers like abaca and jute, imported at rock-bottom cost from overseas, primarily the Philippines. In fact, the Philippines had become such a major exporter of these bast fibers that they became known domestically as "Manila hemp," leading many to confuse them with much stronger cannabis. Not all were fooled, as at least one observer noted that "abaca and jute, inferior to cannabis in almost every way, have only one advantage to recommend them: they are very cheap."

Bernagozzi's scavezzetrice raised the possibility of revitalizing the American hemp economy at a fortuitous time, when a fast-growing new idea would consider taking hemp farming as one of its causes. The early American environmentalist movement, taking inspiration from John Muir's travels and leadership from Theodore Roosevelt's bully pulpit, had begun to aggressively promote the idea that America's wilderness was a valuable resource worth protecting. Of particular interest was paper; new chemical pulping processes had opened up vast swaths of forest to America's paper mills, creating easy access to literature and the news but also threatening to destroy the country's old growth forests.

In this context, the USDA released a 1916 pamphlet identifying a neat solution to the problem, one which would secure an abundant supply of high quality paper while simultaneously protecting America's forests.[292] It would do this by switching paper production from trees to a different plant - one which grew very fast, which produced an abundance of useful biomass per acre, and which would not have to take the place of food or cash crops. Taking seriously their mandate to invest in America's farmers, the Ag Department hired an eminent agronomist to a full-time position devoted to studying the plant and developing uses for it. That agronomist's reports would astound them.

Lyster H. Dewey stopped to wipe the sweat of his brow. The summer days could get really hot in Arlington, Virginia, where the fastidious gentleman farmed the same plot for 48 years. The parcel, known as Arlington Farms, had been the site of Dewey's daily labors since 1896, when Dewey first went to work for the U.S. Department of Agriculture. It had been a plum assignment for the plant lover, but nothing had been so exciting to Dewey as the fantastic results he had witnessed while growing hemp at Arlington Farms in the early 1900s, all under the auspices of the federal government.

Through careful breeding of some of the most famed cultivars in the world, Dewey had developed new strains of cannabis land races which augured to overcome one of the most vexatious roadblocks to the coming cannabis renaissance: the length of the plant's fibers. Ironically, most industrial cannabis cultivars grown in the U.S. before 1900 had been deliberately bred to grow very long bast, a trait which generally strengthens the end product sail or rope. But this useful genetic trait actually hindered cannabis's resurgence in the 20th century, when no domestic facilities existed which could process fibers twenty feet long. Meanwhile, a minority of U.S. industrial strains were grown for their seed, a commodity traditionally favored as a cheap and nutritious bird seed.[293] Through careful breeding, Dewey had developed "dual-use" cultivars which solved two problems at the same time. These new strains, crossing Chinese imports[294] with local Kentucky cultivars, could grow both useful fibers and valuable seed on the same plant, creating a double payout which instantly made cannabis more viable as a cash crop. Even more significantly, these dual-use cultivars grew to heights similar to flax, an already-popular fiber crop with an established domestic infrastructure. Thus, it appeared likely that bast fibers from these new strains could be used in flax-processing facilities with minimal adaptation, removing a major roadblock to resurgence.[295]

Yet for all that Dewey still had not found the end of the plant's resourcefulness, and when the seasoned scientist began to understand the plant's extraordinary potential as a source of paper, he gave the matter special attention. Dewey had served on the team which had first identified cannabis as a superior source of paper in 1916, finding that an acre of hemp could sustainably produce four times the paper as an equivalent acre of forest by utilizing the plant's woody hurds, which could be found abundantly wherever hemp was already grown. For centuries, these rigid columns in the plant's interior had been treated as mere discards: after the poor farmer or hardy slave broke the cannabis fibers free from the woody core, the broken bits of hurds were cast aside. Some farmers mulched it for animal bedding, others burned it as a fuel, but almost no one seemed to realize the resource's full potential. Examining this common "waste material," Dewey discovered that it contained 30-45% cellulose - an unusually rich source (the plant's bast fibers contained even more - up to 77%). The green-thumb bureaucrat realized that he had found a sustainable way to produce paper without felling a single tree.

The USDA's 1916 proposal brought the U.S. in line with history, as both the first Gutenberg Bible and the first draft of the Declaration of Independence had been set down on paper made of wet cannabis bast fibers pressed together and dried, a process long celebrated for producing an exceptional, archival-quality product. Determining that this costly process probably wouldn't benefit the cannabis farmer, Dewey instead turned to the hurds and asked, what if all that cellulose could be made into paper, too?

Working with existing paper mills, Dewey ran a series of experiments processing cannabis hurds into paper using minimal adjustments to the existing facilities, and found that a product usually thought of as waste could indeed be cheaply processed into paper generally comparable to the ubiquitous wood pulp sheets then in use.[296] Hurd-based paper was nowhere near as exceptional as paper derived the old-fashioned way from hemp bast fibers, but it was still quite good enough for a host of applications, most notably for newsprint. Successful experiments in re-tooling wood pulp paper mills to cannabis production suggested that a third market had been created for the American cannabis farmer, who could now sell fibers, seeds, and hurds - almost the entire plant. These advances presaged a much-needed new income source for the American farmer and a more efficient use of resources - nothing less than a revolution in America's agrarian economy.[297]

Any other self-respecting environmentalist might have hung up his hat at that point, but the driven Dewey discovered even more ecological advantages of hemp. Writing for the USDA in 1913,[298] he summarized his discoveries of the plant's properties which promised to improve not only the farmer's land but also his ledger balance. For example, the deep and luxurious spread of the annual's root system could penetrate deep enough to draw up nutrients other crops couldn't access, leaving behind a deeper, airier, richer soil when it died in the fall. The rotting root system creates ideal conditions for the spread of mycelium and earthworms - two beneficial support species - and the plant's falling leaves create an effective mulch to lock nutrients in and protect beneficial ecosystems. While the leaves remain on the plant they provide a different benefit, because the thick canopy of the field blocks out all light to any would-be competitor; Dewey identified only two North American weeds (bindweed and morning glory) which had any chance of growing quickly enough to compete with hemp's celerity. And since hemp was found to have a natural resistance to most North American insects as well, Dewey confidently reported that the hemp farmer would need few if any chemical herb- or pesticides.[299] While the consummate scientist acknowledged that growing hemp in monoculture would eventually deplete the soil of vital nutrients, he solved this problem by designing a number of crop rotations which utilized cannabis in conjunction with other crops to increase the depth, loam, and richness of the topsoil over time, while saving the farmer money in the process. Dewey's corner of the USDA became nothing less than a font of hemp evangelism, proudly preaching the gospel an impassioned horticulturalist believed would save his country.

The titanic Hearst media empire had other plans. From the vantage of William Randolph Hearst, possibly the most powerful man in America, an economic boon for America's farmers potentially meant a major blow to his bottom line. The man whose life had inspired Citizen Kane had sunk no small portion of his fortune into purchasing vast tracts of forest in California with the specific intention of feeding the paper mills and presses of a massive network of 28 major U.S. newspapers; but after Dewey's discovery of 1906, he suddenly found himself beleaguered by a new group of environmentalists with President Roosevelt at their head, cheekily insisting that not all forests should be cut down. Pressure built when a Jewish immigrant named George Schlicten approached the rival Scripps news service with a plan to economically convert their newsprint production to hemp-based paper, based around his improvement on the Bernagozzi design.[300] Schlicten claimed that his machine could help Scripps produce newsprint at less than half the cost of wood pulp paper, and when the costs of reforestation were factored in, he was probably right.[301] But the coup de grace came from none other than President Roosevelt himself, who designated Hearst's California forest as protected land before Hearst had time to truck in the saws. Hearst was incensed. After many failed attempts to overturn the decision, the newspaper magnate went south, finding a more willing partner in General Porfiero Díaz, the thuggish dictator of Mexico,[302] who presided over the purchase of a 1.6 million acre tract of Mexican forest with the blessing of a government not subject to the plebiscites of environmentally-conscious citizens.

Or so Hearst thought. When José Doroteo Arango Arambula heard that a rich gringo from the north had made a deal with the corrupt Díaz to clear-cut his beloved Baricoa forest, the petty horse thief and murderer decided it was time to make something more of his life. To stand for something, and against the long deplorable oppression of his people. To take action.

Rounding up a posse of his close friends and other trusted men, José led an expedition into the heart of Baricoa, where they employed a clever mix of guerilla tactics to hold off any attempt secure the forest. Hearst, facing a second devastating blow to his balance sheet, put intense pressure on Díaz to arrest the ragtag gang; Díaz complied, sending in his army in 1910 to take the forest back.

Unknown to Hearst, Mexico had just crossed the tipping point. Thousands of Mexican peasants, trodden down through long years of abuse, drew inspiration from José's occupation, eagerly joining with a group actively taking up arms against oppression. In short order, the situation was entirely out of hand: Hearst and Díaz stood powerless to stop the swelling new movement who had taken José Doroteo Arango Arambula as their leader, renaming him "Pancho Villa." Worse yet (for Hearst), the Villistas enjoyed broad popular support among Americans, who felt sympathy for Villa's fight against an undemocratic regime and pressured President Taft into providing his army with arms, ammunition, and training. As Díaz's hold on Mexico grew more and more tenuous, Hearst began to watch his valuable forest slip through his hands. Seeking a chink in Villa's public relations bonanza, Hearst dispatched a correspondent to embed with Villa's army and report back on any unseemly facts which could be used to smear the Mexican revolutionary.

The fight for revolution dragged on for years. President Woodrow Wilson, succeeding Taft, desired a happy ending and threw his weight behind Venustiano Carranza, who had by 1915 begun to consolidate power as the newly appointed President. Pancho Villa, regarding Carranza as just another dictator, fought on; but soon found that his supply of American armaments had dried up. Feeling betrayed by his former American friends, Villa (or possibly a lieutenant on his own initiative) took matters into his own hands and attacked an ammunition depot in Columbus, New Mexico, making off with the weapons and supplies he badly needed.[303]

The decision proved shortsighted. The outcry against the man the American public once viewed as a revolutionary hero now brazenly invading the United States became deafening. Hearst sensed his advantage. His correspondent, upon arriving in Villa's camp, reported that the Villistas had taken to singing a particular song, used as a rallying cry of sorts to boost their morale. The words were:

La cucaracha, la cucaracha,
Ya no puede caminar
Porque no tiene, porque no tiene
Marijuana que fumar.

(Translation:
"The cockroach, the cockroach,
Already he cannot walk,
Because he doesn't have, because he doesn't have
Any 'Marihuana' to smoke")

The song came with the practice it lampooned, tracing the spread of the Mexican joint from its origins in the country's interior to the very edge of Texas, where the custom had just begun to appear at the same time that Pancho Villa began to attain some notoriety.[304]

In border towns like Laredo and El Paso, Texas, local prohibition ordinances followed closely behind - the result not of any scientific inquiry but rather of the cursory copying of a foreign policy in the face of an unknown but supposed evil.[305] Since the mid-19th century, Mexico City's myriad newspapers condemned *mariguana* in the harshest terms and called for its criminalization, apparently unaware that it had already been banned - as *pipilzintzintlis* - nearly a century prior. If such calls were significantly opposed in any way, no record of the struggle survives; by the late 19th century, a consensus had arisen in favor of prohibition. The first Mexican government to ban *mariguana*'s sale was Mexico City in 1869; the state of Oaxaca followed suit in 1882; in 1888 the city of Cosala in the mountainous Sinaloa district went even further, banning both the herb's sale and its use. By 1891, the entirety of Mexico State had criminalized.[306] These and other measures passed virtually unopposed.

Having no familiarity with the drug or its effects but empathizing with the racist scorn and paranoia which accompanied its prohibition just across the border, the town of El Paso simply continued the trend, passing what may have been the first cannabis prohibition in America (1914),[307] roughly three hundred years after Jamestown Colony levied a fine on any colonist who didn't grow enough. The arrival of the Mexican custom, and the prohibitions which followed it, fatefully coincided with the paradigmatic King laws in Canada that prohibited opium; and so in almost the same moment, careerist American legislators found inspiration from both North and South, in pioneering legislation which used the excuse of drug use to justify and enforce a racially-stratified society.

The crux of ignorance and punitive precedent granted Hearst a narrow window of opportunity, and the magnate intended to seize it. American public opinion of the Mexican people teetered on a tipping point in the late teens, with many businesses commenting favorably on the industriousness of the people and the capitalist-friendly Diaz regime but others wavering with uncertainty of foreign ways. If the man behind the nation's largest mouthpiece could stir that uncertainty into fear, he could turn the tide of public opinion against the Villistas and restore the Diaz regime which had promised him a mountain of pulp.[308]

But the window was closing fast. As early as April 1917 the U.S.D.A. had uncovered the connection between *mariguana*, hemp, and extract of cannabis indica;[309] if Hearst were to exploit the temporary confusion to multiply the American fear of the unknown, he would have to act quickly. Soon every paper in the Hearst empire began running a variation on the same scenario often related about cocaine or heroin: a "decent" white woman getting seduced and/or raped by a Mexican male turned suddenly insane and violent by a depraved drug he kept ready at hand.[310] Newspapers all throughout the Hearst chain took pains to associate the drug's use with Mexicans in general but Villa in particular, seeking to build on the drug's connection to the man who had exhibited bold violence against the U.S.'s sovereign territory. The story was entirely false - the tee-totalling Villa disfavored use of all intoxicants, and 'La Cucaracha' was probably a slight aimed at Villa's rival Victoriano Huerta, popularly known as "El Presidente Mota" or simply "El Marihuano" for his supposed laxity toward its use in his army.[311] Hearst, rarely one to let an inconvenient truth dilute a sensational story, ran with it anyway.

Instead of reporting Mexican adoption of "cannabis indica," Hearst preferred the unfamiliar term "marihuana": if Americans learned that the "new drug" came from the same plant as did a medicine still common in the medicine cabinets of older generations and even the fiber hemp still often found in the Midwest hinterlands, they would not have felt sufficiently threatened for Hearst's purposes.[312] So, rather than use the familiar terms "hemp," "Indian hemp," or "cannabis," Hearst cloaked the drug's domestic identity with an exotic *nom de plume*.[313]

The strategy worked; California, Utah, Colorado, Texas, Iowa, New Mexico, Arkansas, Nevada, Oregon, Washington, Idaho, Kansas, Montana, Nebraska, Wyoming, North and South Dakota, and Oklahoma all passed anti-cannabis legislation between 1913 and 1933, with the majority following the King model of drawing a dividing line between recreational and medical use.[314] The public which had recently expressed strong sympathy for the Mexican revolution had become willing to lock up Mexican refugees en masse, copying the campaign to control Chinese immigration through opium legislation and making cannabis the collateral damage in a war for racial supremacy. In most cases, mere mention of the weed's origins as a "Mexican drug" sufficed to ensure breezy passage of its prohibition.[315]

This rapid wave of bans was made possible in part by a hysterical impulse raging through the twenties in reaction to the rise of sophisticated organized crime networks which sprang up in America after 1919 - the year when the federal government clamped dramatically down on the drug habits of Americans through the passage of the Volstead Act and the Supreme Court's pair of decisions upholding the Harrison Act's penal enforcement regime. Right on the eve of the "Roaring Twenties," all legal trades of morphine, heroin, cocaine, and alcohol simultaneously vanished, kicking every addicted American to the curb and leaving a very large niche. In a capitalist country, it was inevitable that someone would fill it.[316]

The Twenties were the heyday of Lucky Luciano and Al Capone, but the original "OG" was Arnold Rothstein, a mathematical genius who masterminded the most audacious international smuggling ring since the Opium Wars; his pioneering blueprint for the drug-smuggling cartel has been used with only few modifications ever since. That was typical

Rothstein - he also invented the first multi-million dollar bond ripoff - but perhaps the best idea he ever had (in terms of making Arnold Rothstein richer, the only cause he cared about) was smuggling booze. Almost as soon as the ink was dry on the Volstead Act, "The Brain" had a sophisticated international syndicate up and running, with its tendrils equally deep in the rum warehouses of Europe and in the wallets of local law enforcement. By the early twenties Rothstein sat on top of a massive and diversified criminal enterprise, running illegal gambling, blackmail, bond scams, and of course liquor.

He had always shown promise. Born to a respectable Orthodox father who rubbed elbows with the likes of Louis Brandeis, he excelled in math, high stakes cards, and smoking opium.[317] Being slight of build himself, he teamed up with notorious Gotham muscle Monk Eastman, who always effectively collected whatever credit the diminutive Rothstein lent out at high interest. But he really hit the big time when he realized that the city's insular network of low-paid messenger boys carried millions of dollars of Liberty bonds in their knapsacks every day - and that certain of these boys could be trusted to falsely report armed robberies in exchange for a cut of the take. The take itself was brilliantly monetized through multi-party collateral transactions which left no paper trail to incriminate their designer - if anyone ever went to jail, it wasn't Rothstein. After the lid blew open on the illegal fixing of the 1919 World Series, everyone assumed Rothstein was behind it; and while he never faced indictment for the crime, he never tried too hard to dispel the rumor, either.

Yet it was booze that made Rothstein's fortune, and the man himself a target. His plan for that was simple: through his contacts on the Continent Rothstein imported boatloads of liquor to the Eastern seaboard, where appropriately calibrated bribes assured easy entry. From there the alcohol spread out to pre-established distribution networks in places like Chicago, Philadelphia, or St. Louis and trickled through the hierarchy down to the speakeasies, who never knew the identity of the ultimate mastermind. In this plan he was greatly aided by the federal agents of the Prohibition unit assigned to catch him; because the majority of these agents were political appointees with no prior detective experience, their incompetence quickly became a matter of national embarrassment.

It was the simplicity of the plan which, more than anything, drew the competition. Some, like Chicago's Al Capone, set up sophisticated distribution networks of their own; but nearly all of the players who swooped in to sate the American demand for liquor sooner or later turned to violence, an aspect of the trade most infamously manifested by the Valentine's Day Massacre of 1929 which resulted in the execution-style murders of seven alleged gang associates. Long before that notorious day, Arnold Rothstein had gotten out of the violence-prone racket, following a business philosophy he summed up thus: "I wasn't fifteen years old before I learned my limitations. I never played with a man I wasn't sure I couldn't beat. I knew how to size them up. I still do. That is all there is to making money."[318]

Watching the federal government pass ever more restrictive legislation on narcotics, Rothstein sized up Uncle Sam and decided to play. After the 1924 Porter Act outlawed all domestic manufacture of heroin, the temptation became simply too great to resist. Government estimates had put the number of domestic opiate addicts at anywhere between a hundred thousand and a million (the lower number is probably more accurate), and with all maintenance clinics shut down by federal law,[319] these addicts had nowhere to go to satisfy their cravings except the street dealer. Brazenly sending agents to Germany, the Netherlands, or France to purchase heroin by the ton (no questions asked), Rothstein smuggled it to the U.S. using the same methods and distributed it through the same network as he used for running booze.[320] Only this time, he kept a low profile, so that almost no one noticed when he quietly cornered the loyalty of the largest gang of street drug dealers in America.

All the average American knew was that crime rose sharply in the twenties, fueled on the one hand by violent acts perpetrated by brazen smugglers of booze and on the other by the petty thefts of some holdover from heroin's heyday who could no longer afford Rothstein's rake. Perversely, these developments actually vindicated the anti-maintenance lobby, which finally witnessed proof of their long-professed theories about the link between narcotics and crime; the fact that the crime wave was caused by the very laws they supported was conveniently overlooked. With such a strong association between drugs and crime burned into the public mindset, the claims of Hearst newspapers - that a new vicious drug called "marihuana" had arrived from the south, whipping Mexican immigrants who carried it into violent thugs - slipped easily into the cultural milieu.[321] By the dawn of the 1930s cannabis had been hemmed by restrictive legislation in the northeast and outright bans in the southwest.

Yet, for all the boldness of Hearst's success, none of it actually solved his problem: he still couldn't cut down his forest in California, and he certainly couldn't cut down his forest in Mexico while a civil war ravaged the countryside. None of the states which had responded to his "Mexican scare" campaign had been big hemp producers; the bulk of U.S. production had centered in places like Wisconsin and Kentucky.[322] And by the 1920s the stakes for Hearst had been raised considerably, as a brilliant group of fringe scientists began to prove that even the pioneering work of Lyster H. Dewey had only begun to plumb the astonishing depths of the plant's potential. Calling themselves "chemurgists," this gaggle of organic chemists set their sights on an audacious goal: to render petroleum obsolete.[323]

The chemurgists had already developed thousands of new applications derived from hemp products between 1900 and 1938, when a *Popular Mechanics* article noted that

> [h]emp is the standard fiber of the world. It has great tensile strength and durability. It is used to produce more than 5,000 textile products, ranging from rope to fine laces, and the woody 'hurds' remaining after the fiber has been removed contain more than seventy-seven per cent cellulose, and can be used to produce more than 25,000 products, ranging from dynamite to Cellophane.[324]

None of the chemurgists pursued this goal more passionately than Henry Ford. The engineering and logistical genius who made an indelible mark on American industry held dearly to many strange ideas, such as his belief that a businessman who succeeds with the support of his community owes that community an obligation of largesse, or that every factory worker on the Ford line should be able to afford his own Model T. Yet, through the lens of history, none of his ideas may seem stranger than his early plan to revolutionize the automotive industry he pioneered with a version of a popular sedan designed to never need a drop of gasoline. Ford, foreseeing a time when the U.S. may not be able to produce as much petroleum as it consumes, began looking for a chance to create a permanently renewable domestic energy supply while transferring profits from large oil companies to small family farms. He found what he considered the perfect solution in the cellulose-rich hurds of the hemp plant, which he used as feedstock for huge vats of fermenting ethanol, producing an automotive fuel with a higher octane rating than gasoline and could be grown right out of the ground, year after year.[325]

By the 1930s, these and countless other chemurgist developments had begun to pose a threat to powerful petrochemical interests; and one concern, the massive DuPont petrochemical company, stood to lose more than any other. Founded in 1802 by E.I. DuPont, an immigrant chemist fleeing the French Revolution, the company's gunpowder manufacturing operation outside Wilmington, Delaware turned a handsome profit in the nineteenth century, making a killing during the War of 1812 and later the Civil War, when the company became the leading supplier of the Union Army's artillery needs. But while wars always provide stable returns on investment, around the turn of the 20th century the company began trying to diversify its revenue stream. One comical artifact which still exists from this period is a 1910 pamphlet titled "Farming with Dynamite"; but when it became clear that America's farmers failed to line up for the chance to blow their fields to smithereens, the company realized that it would need to learn how to make more than just explosives.

Thus, around the turn of the century, the company set up the DuPont Central Research laboratory, which with the help of eminent chemist Wallace Carothers went on to discover an impressive list of new petrochemical-based materials, including polyester, neoprene, and Cellophane. While these investments in material sciences undoubtedly sufficed to make a rescendant hemp industry a financial threat, the DuPont company had even more unique circumstances to force focus of its ire on a previously harmless plant. In 1914, for example, the DuPont company invested $25 million in the struggling General Motors, part of a deal which resulted in the election of Pierre DuPont as GM's President and Chairman of the Board; this move put DuPont squarely in competition with Henry Ford and his crazy ideas about growing fuel from the soil. Even more endangered by the looming hemp renaissance was DuPont's lucrative business selling chemical pulping cocktails to paper manufacturers like Kimberly-Clark, which required the chemicals to break down the gluey lignins in wood before they could make paper; because of the low lignin content of hemp hurds, paper made from hemp would require less than a quarter of the harsh chemicals, virtually destroying a valuable revenue stream.[326] Yet the greatest threat to DuPont's bottom line was a venerable competitor to a new petroleum-based material which the company had expended considerable sums to develop, on an audacious bet that it would cement DuPont's place as the world's preeminent chemical conglomerate: Nylon.

At the time of Nylon's invention in the thirties, the cost of petroleum was so low that the petrochemical cocktail could easily compete with even the low-quality rope fibers imported from the Philippines; but an abundance of hemp-derived products on the market had the potential to change the calculus. If a hemp farmer could simultaneously sell both the bast fibers used to make rope and the hurds used by chemurgists and paper-makers for their products, he could theoretically unload each commodity at a lower price than he otherwise could; by sharing production costs across multiple products, he could still make a profit. DuPont couldn't risk allowing such stiff competition from an ultra-strong fiber which could be produced dirt cheap; but the murder of Arnold Rothstein presented a rare opportunity to unlevel the playing field.

After Rothstein's death, the lid finally blew off his dealings and the full extent of his complex criminal empire came into focus. Details came to light despite the determined ignorance of the NYPD when U.S. Attorney Charles Tuttle (a Republican party man who despised the city's Democratic administration) launched his own investigation, ordering his men to break into a room in an unnamed Manhattan hotel; they found $2 million worth of opium, morphine, cocaine, and heroin inside. NYPD learned about the raid from the newspaper coverage the next day, in which a triumphant Tuttle announced "the single biggest raid on a narcotic ring in the history of this country," pioneering the hyperbolic style still used in drug enforcement press releases.[327]

But the shame and shock the public felt over the fruits of Tuttle's investigations peaked even higher after the crusading attorney uncovered revelations of a decade-long legacy of corruption and scandal. Such shenanigans were suspected from the corrupt Democrats of Rothstein's home town; for decades Tammany Hall had striven to set the gold standard for cronyism, patronage, and quid pro quo. They usually succeeded. But the awe-inspiring audacity and breadth of Rothstein's operation couldn't hide behind a mere gaggle of greased palms in the Big Apple; a narcotic distribution network as big as The Brain's could only continue with the cooperation of big government at its worst. (Or, as a lawyer for the Rothstein estate once put it, "If the Rothstein papers are ever made public, there is going to be a lot of suicides.") When Tuttle convened a grand jury to investigate whether Rothstein had compromised even Colonel Levi Nutt, Commissioner of the Federal Narcotics Division of the Prohibition Unit,[328] a complacent Congress was finally forced into action.

One of the first Representatives to approach Tuttle was Stephen G. Porter, author of the 1924 Porter Act banning domestic heroin manufacture and the most vocal proponent for dedicated addiction treatment facilities since Fitz Hugh Ludlow.[329] (In a twist of cold irony, however, Porter completely eschewed Ludlow's linchpin prescription: the palliative use of cannabis to ease the patient's hellish push through withdrawal.) Using Tuttle's scandalous grand jury report as ammunition, Congressman Porter pushed ardently for his next legislative cause, a shakeup of the Treasury's bureaucracy. Indeed, Porter had argued for such a change long before the corruption at Narcotics had been revealed, believing that effective narcotic control could not be attained without cooperation between the Treasury and State Departments, which could coordinate a single U.S. policy at the domestic and international level, simultaneously. Porter took advantage of the Nutt scandal to push for a new Federal Bureau of Narcotics, separate and autonomous from the Prohibition Unit, which could coordinate directly with State.[330] To lead this new Bureau in Nutt's place (the Colonel had been demoted to field supervisor and banished to the frozen wastelands of Syracuse, New York[331]), Porter advocated strongly for a civil servant who exhibited equal ease with police work as with diplomacy. Porter scored big when Congress approved his proposed Federal Bureau of Narcotics in March of 1930.[332]

To head the new Bureau, Congressman Porter strongly supported Rear

Admiral Mark L. Bristol, a disciplined investigator with proven negotiation skills. Bristol, however, was not the only one who vied for the job; many an aspiring public servant chomped at the bit while pondering the chance to set the mold for an important new government agency. Among the better-qualified and even better-connected candidates was Elizabeth Washburn Wright, widow of the late Dr. Hamilton Wright (Dr. Wright, after his failure to secure a treaty at the Hague conventions, watched his fame and public esteem plummet; he died in 1917). Mrs. Wright had been no passive bystander of her husband's diplomatic efforts, and by the late 20s she had become known as a seasoned crusader of international drug policy, equaling her husband in enthusiasm and surpassing him in tact. While the widespread chauvinism of the period probably precluded a woman from filling the role of "top cop," Wright's diplomatic credentials could not be denied, and she earnestly dreamed of smashing glass ceilings.

Thus, when Wright became a regular soothing presence at the bedside of the ailing Porter, whose cancer diagnosis confined him to a hospital bed in 1930, it may not have been charity alone which motivated her ministrations. As the scion of a political dynasty (her father and four of her uncles had all served terms in Congress), the charming Mrs. Wright had been weaned directly from her mother's breast to Congressional politics; she had the ear of numerous influential Republicans at a time when the GOP controlled both chambers and the White House.[333] Her influence and kindness toward Porter paid off when he appointed her his envoy on Capitol Hill, where she whipped up the necessary votes for his autonomous Bureau. After the bill's passage, he asked her to lobby for his choice of Admiral Bristol for the Commissioner position, which she dutifully did; but always expecting a high-ranking position of her own, from which she might get the promotion to Commissioner under a more enlightened administration, perhaps not that far off.

It seemed a real enough possibility. Mrs. Wright's charm and political legerdemain had paid off several times before, and most of the Treasury officials she set out to persuade came away with an impression of her as capable and convincing.[334] One official who met with Wright in early 1930, however, hid behind his smile of welcome the wary watchfulness of a canny rival. This young man, as handsome as Wright was beautiful, had charms of his own and an innate sense for the workings of bureaucracy; he had recently landed the biggest break of his bureaucratic career when he had been selected to succeed the disgraced Nutt on an interim basis by his political benefactor. Sensing in Wright a dangerous level of skill, he made plans to neutralize her. Wright, too, sensed the presence of a rival, and diplomatically wrote to President Hoover's secretary that she had found the interim commissioner to be a "very alert and intelligent man," with the potential to be "a most valuable lieutenant or assistant to the Admiral."[335]

But the cards were already stacked against her, though she didn't know it yet. When Stephen Porter finally succumbed to cancer on June 27th, 1930,[336] his brainchild the Federal Bureau of Narcotics had still not been created and its commissioner had still not been named;[337] he may have died believing his choice of Admiral Bristol would be posthumously honored. Treasury Secretary and fellow Republican Andrew Mellon did indeed pay a measure of respect to Porter's exemplary crusade against drug addiction by waiting until the day after his funeral to name the Bureau's new commissioner, the handsome rival Elizabeth Washburn Wright had met in Washington that spring. This young man, who had exhibited some early success in the Prohibition Unit, had to Mellon the added advantage of being absolutely dependable when it came to following orders: his name was Henry J. Anslinger, Jr., and he was married to Mellon's niece, Martha Denniston.[338]

It was Martha's influence which adroitly transferred Anslinger out of his assignment in a remote Venezuela backwater to a cushy appointment in the Bahamas, and her cozy relationship with the head of the Prohibition effort ensured that her husband would rise quickly through the ranks. It is indeed doubtful that Mellon paid any attention at all to the impassioned drug crusader or dying Congressman who wished to see the Narcotics Bureau created by their designs; he had his own reasons - most of them financial[339] - for his choice, which he had probably made years before. When Mellon made his recommendation to President Hoover, he received strong support from the Hearst newspaper chain, which Anslinger later credited as crucial to ensuring a swift and smooth confirmation.[340]

Anslinger, who at the time of his selection to run the Narcotics Bureau had scant experience fighting the narcotics traffic, possessed police skills just north of middling, diplomatic skills of considerably greater worth, and a genius for bureaucratic brinkmanship which rivaled J. Edgar Hoover's.[341] He exhibited this last skill early on by his deft handling of Elizabeth Wright: he offered her a job, but only on the condition that she could never do any policy work. Since policy had been the focus of her entire career, the humiliated Wright indignantly declined.[342] Yet Anslinger placated her shortly thereafter by arranging for her selection to a blue-ribbon government commission to study the opium problem in Philippines, thus creating an opportunity for her to save face and simultaneously ensuring that one of his greatest rivals remained on the other side of the world while he created his precious Bureau in his own image.

And that image was punitive. From the beginning of his career in public service to its bitter end, Anslinger never wavered in his belief that any government policy could engender the awe and respect of the criminal classes if only it were enforced with the threat of sufficient jail time. [343] From the very first days of his tenure as Narcotics Commissioner, he used every cent of his meager appropriations[344] to aggressively prosecute Harrison Act violators, exorcising, perhaps, the haunting monsters of childhood memory. His "tough on crime" stance met the approval and even active encouragement of presidents across the political spectrum: after Franklin Delano Roosevelt ejected Hoover from the White House, one of Roosevelt's close advisers stopped by Anslinger's office to let him know that any appeal to the president for clemency of a drug dealer might as well have Anslinger's resignation attached, because that's how it would be received.[345]

The Hearst chain which had helped to secure his appointment immediately agitated for federal anti-cannabis laws, and many western states demanded action. Beginning in the late twenties, the Narcotics Bureau began to receive insistent requests to criminalize from western politicians, especially from Colorado, where migrant beet field workers had spread the practice of smoking joints around the countryside.

Yet Anslinger's answer to such prayers never wavered: averring that the marihuana menace was a matter for for states to solve themselves, he urged the supplicant to contact their state representatives to support the so-called Uniform Narcotics Acts which had begun to trickle out of statehouses in the thirties. While he always agreed that marihuana was a terrible problem,[346] Anslinger didn't even attempt to do anything to solve it, even if the step could mean more money from Congress to fight a new evil.[347]

It was a response constrained by a popular insistence on limited government. The federal Constitution, as interpreted by the Supreme Court of the 1930s, had never before authorized Congress or the Washington bureaucracy to interfere with domestic crops - a power then well understood to be reserved to the states. Although every indication exists that Anslinger wanted badly to criminalize marijuana every way he could, the Treasury's lawyers kept the drug warrior in check until the constitutional issues could be sorted out.[348]

In Mexico, federal cannabis prohibition only became possible after ratification of an entirely new constitution in 1917;[349] in the wake of a new, centralized federal government Mexico passed the first ban of its kind in 1920, and made *mariguana* illegal throughout the land.[350]

In the U.S. of the early thirties, when the country thrilled with the long unwinding of the knot of Prohibition, a new amendment authorizing bans on cannabis appeared unlikely. The FBN owed its entire reason for existence under the 1914 Harrison Act, which the Supreme Court had upheld by a tenuous 5-4 margin only 14 years before. To set any course which might subject the Harrison Act to another round of judicial review might be the bureaucratic equivalent of Russian Roulette; the Supreme Court could wipe the entire project away with five strokes of a pen. Thus Anslinger, under the advice of FBN counsel A.L. Tennyson and Treasury lawyer Herman Oliphant, steered well clear of any policy which might be viewed as an enlargement of federal power under the Harrison Act.[351]

Without any doubt, any attempt to ban cannabis on the federal level would have fit that description. The slim Supreme Court majority upholding Harrison found it persuasive that cocaine and opiates overwhelmingly arrived in the U.S. from overseas, thus coming under Congress's power to regulate international commerce. Consequently, federal prosecutors could claim that anyone found in possession of any of these drugs was presumptively in violation of federal law if they didn't have proof of their registration - a nifty loophole in the Constitution. Marihuana, by contrast, could be found growing on the sides of roads and in abandoned lots in dozens of states throughout the Union, so the legal theory of presumptive illegal importation would have no logical basis. Trying to regulate cannabis under the Harrison Act ran a significant risk of drawing rebuke from the Supreme Court.

None of this meant that Anslinger was soft on marijuana; the Commissioner had his own plan, which he believed stood a much stronger chance of surviving the Supreme Court's scrutiny. On the federal side, he advocated for one law banning the importation and exportation of cannabis, and another banning its sale across state lines. On the state side, he urged a cooperative policy among all 48 state legislatures in which each state adopts legislation banning the cultivation, sale, and even possession of cannabis under its general police power. With cannabis locked down on the local, national, and international levels simultaneously, Anslinger's proposal was pioneering in its punitiveness and unprecedented in its reach, even if its design was constitutionally cautious.

That's why Anslinger's first major duty as federal Narcotics Commissioner was to try to shape state policy.[352] In fact, a model state bill similar to the one he advocated for had already begun to come together, under the leadership of the American Medical Association's Dr. William Woodward, a well-respected physician who had also been trained as a lawyer.[353] Calling his policy the Uniform Narcotic Drugs Act, Dr. Woodward had worked within his Association since 1925, seeking to build support and consensus for the law among his peers.[354] By 1930, that effort had borne some fruit, creating draft legislation which would ban or severely restrict many of America's most dangerous drugs, like cocaine or heroin. Dr. Woodward's proposal gained great support despite a distaste within the medical community for meddlesome government regulation; the community had finally owned up to the dangers of opiates and cocaine, recognizing that the crisis in public health required effective restrictions. In any case, since a mishmash of narcotics laws had already been passed by numerous states, the AMA would rather deal with one simple, unified rule than a patchwork of policies from sea to sea.

Dr. Woodward's proposed language included criminal penalties for possession of heroin, morphine, and cocaine - and a provision called "Subparagraph Twelve" for banning cannabis. His committee listed cannabis not from any national problem but rather a general assumption that it was a "narcotic" - a designation given at the time to any drug which caused either soporific or euphoric effects. Yet even with the doctor's patient prodding, four years of discussions passed by without producing a model law.

Anslinger's efforts breathed new life into the project. Right out of the gate, the Commissioner took a keen interest in

> [s]ub-paragraph Twelve covering cannabis sativa and its derivatives [which] does not at present come within the purview of the two principal Federal narcotic laws. However, many complaints have reached this office of the abuse of this form of drug in certain localities. In view of what is understood to be its very limited medical use and its lack of dependability as to potency, query is made whether this drug could not be absolutely proscribed and its limited function as a medicine met by some substitute. From the limited information in possession of this office it is believed that the elimination of the abuse of this drug is a consideration which greatly outweighs that of its possibly very limited medical use, particularly if the latter need could be met by some less potentially harmful substitute.[355]

While the commissioners did not seem unduly offended by the bureaucrat's interjection, Anslinger's proposal still met resistance - ironically, from Dr. William Woodward, who had proposed its banning in the first place. In 1929, desiring to better understand the subject matter, Woodward sent out surveys to pharmacists throughout the United States, asking them for firsthand data on American cannabis usage; and the responses he got surprised him. Of all the pharmacists surveyed, not a single one reported any known incidence of abuse;[356] all but one disputed the notion that it was "habit-forming." Most intriguingly, more than a few respondents surmised that government bans might actually cause abuse to arise where previously none existed, merely by operation of publicity.[357] Quite a few referred Dr. Woodward to the results of a 1925 study conducted by the U.S. Army in the Panama Canal Zone in response to reports that troops stationed there had taken up the habit of smoking joints. Although methodologically crude by modern standards, the study was nonetheless notable as the first time the U.S. government attempted an objective study of the drug's effects on human beings.[358] It was a mistake not to be repeated for nearly fifty years.

The study's final report debunked the insanity thesis, the fear that its use affected troop discipline, and its supposed addictiveness. It concluded, "there is no evidence that marihuana... has any appreciably deleterious influence on the individuals using it."[359]

Dr. Woodward, displaying the skepticism behooving the best of his profession, changed his mind about marijuana once swayed by the evidence. Around the time of Anslinger's first interference with the UNDA's drafting, Dr. Woodward began preparing an alternate draft which began to get strong support from his fellow commissioners. It omitted Subparagraph Twelve's cannabis prohibition and put in its place optional language followed by the following

> Note: Because of the many objections raised to the inclusion of cannabis indica, cannabis americana and cannabis sativa in the general list of habit-forming drugs, no mention is made of them in other sections of this act. The foregoing section is presented in order to meet the apparent demand for some method of preventing the use of such drugs for the production and maintenance of undesirable drug addiction. It may be adopted or rejected, as each state sees fit, without affecting the rest of the act.[360]

Anslinger fought hard, urging Woodward to reconsider his position on marihuana and heavily lobbying Judge Deering, the Commission's chairman, to influence full prohibition. Nevertheless Anslinger lost battle after battle, as subsequent revisions began to favor Dr. Woodward's language over the Bureau's punitive position.[361]

As Anslinger's domineering interjections became more insistent, the relationship between the bureaucrat and the doctor broke down. When Anslinger sent out a form survey to Woodward as part of a general fact-gathering rather late in the process of the Bureau's policy process on marihuana, the lawyer-physician accused the Commissioner of merely seeking to generate political cover over a foregone conclusion. In his survey responses, Woodward doesn't attempt to disguise his condescension:

> I
>
> *What is the quantity of Indian hemp produced in the United States?*
> Ans. - The American Medical Association has no knowledge with respect to this matter.
>
> II
>
> *What is the geographical distribution of the areas where Indian hemp*

> *is grown within the United States?* Ans. - The American Medical Association has no information with respect to this matter except such as may be found in books and journals, with which information, it is presumed, you are already familiar.
>
> III
>
> *What are the medical needs and uses of the drug or drugs produced from Indian hemp?* Ans. - The answer to this question can be found in standard books on pharmacology and therapeutics, which are available in large numbers in the Surgeon General's library in Washington.
>
> IV
>
> *What is the comparative medical value of Indian hemp as domestically produced and as produced in foreign countries?* Ans. - The answer to this question can be found in standard books on pharmacology and therapeutics, which are available in large numbers in the Surgeon General's library in Washington.[362]

With the remainder of the Association, Anslinger fared little better. Soon realizing the inevitable failure of his chosen course, Anslinger took his case to the sensationalist press[363] and awaited a better opportunity.

It soon arrived. In the fall of 1932, just before the Commission's final draft was supposed to be presented, controversy erupted when Dr. William Charles White of the American Pharmaceutical Association pulled rank to complain directly to the AMA's president about Woodward and insinuate that the final preparations should be conducted without him. The AMA's president responded with professed surprise; the APhA had given Woodward its full cooperation in connection with his pharmacist surveys. Woodward, was shocked, too: why, when the end of a seven-year process was finally nearing its end, would these objections suddenly arise? The eminent doctor began to suspect that the brouhaha and subsequent delay had "been deliberately promoted to serve the ulterior purposes of some interest that is unwilling to be known in connection with the matter." [364] Although Woodward never named names, it is entirely possible that Anslinger leveraged the threat of strict enforcement against the pharmaceutical industry in order to cause dissension.[365]

Whether he was behind the discord or not, Anslinger undoubtedly exploited it to his maximum advantage. Elbowing out the temporarily toxic Dr. Woodward, the Bureau Commissioner leaned on Judge Deering to delay finalizing the Uniform Act's final drafting until the dust had cleared.

The Judge agreed, and Anslinger took advantage of the delay to plan his contingencies, drafting multiple versions of the Act to serve his ends.

When the Commission convened for a final time, Anslinger lobbied hard to take the lead, pushing hard for language which would completely prohibit cannabis in all its forms. Yet even with his rival sidelined, Anslinger met stiff resistance, especially from the pharmaceutical lobby, which disagreed with the characterization of cannabis as a dangerous drug. Further, the APhA strenuously objected to the "effort being made by sensationalistic papers to enlarge upon the extended use of marihuana" - it appears that Anslinger's yellow journalism blitz had backfired. When word finally spread around the Commission that Anslinger's proposed language would end not only all drug production, but also the recently revitalized hemp paper, fiber, and chemurgy industries, the punitive version was finally killed.[366]

Dr. Woodward stepped back into the fray, jumping on the chance to rebuild consensus, which in the fall of 1932 appeared near at hand. But the wily bureaucrat had one more trick up his sleeve. At the final meeting of the Commission on October 8th, Anslinger switched tactics, paying respects to the group's decision to keep cannabis prohibition optional yet suggesting a small change to the law as a minor matter of form. Rather than list each affected drug in a separate provision - an unnecessarily cumbersome construction, he argued - why not use general language to prohibit each drug listed in a separate schedule, so that when dangerous new drugs arrive on the scene (as did cocaine and heroin in the late 19th century), legislatures could nimbly react by simply adding the terrifying new drug to the banned list. As a matter of fact, the drug czar just happened to have exactly such language at the ready; no need for time-consuming re-drafts. The council found him persuasive; because no attempt was made at the time to add cannabis to the banned list, the opposition of pharmaceutical interests was neutralized. Dr. William Woodward left the conference with a self-congratulatory aspect, and Harry Anslinger planned how to exploit the loophole he had just engineered.[367]

The stakes for Anslinger peaked around this time, when DuPont's lobbying operation began to ramp into high gear, bringing pressure to bear on Treasury counsel Herman Oliphant to find a way to finish hemp, once and for all.[368] Andrew Mellon intensely lobbied Capitol Hill (and, presumably, his nepotistic avatar at the helm of the FBN). The petrochemical company was getting ready for two major product rollouts which promised not only to increase sales but even to completely remake society after a chemist's dream: new chemical wood pulping and Nylon, both due in 1937.

Anslinger cast his net even wider, taking his 300 federal agents off the beat and embedding them in statehouses to target any politician who might be on the fence about the Uniform Acts, and to push in particular for passage of the Act's "optional" cannabis prohibition.[369] One Bureau employee, Elizabeth Bass of Chicago, attained better lobbying results than almost any other agent by rallying the Women's Christian Temperance Union with narcotic nightmares - an early adoption of the "protect the children" argument.[370] Meanwhile, the Commissioner went back to Hearst and his friends in the sensationalist press, using them to whip up support among the nationwide electorate for the Uniform Acts, setting out to educate the public about marihuana in particular, and urging its inclusion.[371]

Yet even after two years of concerted effort, Anslinger's goal of nationwide pot prohibition remained frustratingly elusive. By spring of 1933 only two states had passed the Uniform Act, and as late as 1935 only ten had passed any version of it, of which three declined to pass the optional cannabis provision.[372] A desperate Anslinger began to feel the pressure of the moment weigh down on him. His hair, previously luxurious and primly parted, began falling out at an alarming rate.[373] His smiling, handsome demeanor began to devolve into the visage of a gruff, bull-faced bureaucrat everyone in Washington feared. He determined to see the government vise clamp down on marihuana.

In late 1934 Anslinger abruptly switched tactics, unleashing an audacious, coordinated media campaign called "The Marihuana Menace."[374] The next two years may have represented Harry Anslinger at his very best - suddenly re-energized and re-inspired, the old sensitive soul began to flower again, taking expression in some of the bureaucrat's best writing, a rapid flurry of magazine articles and press releases finally culminating in his masterpiece, "Marihuana: Assassin of Youth."[375] Forget heroin - this menacing new drug was the worst to ever mar the pure white bread facade of Americana, a narcotic so savage in its effects as to ensure that any person who tried it, no matter how otherwise innocent, would descend surely into madness and death.[376]

He pulled each story from a special file cabinet hc kept in his D.C. offices nicknamed "the gore file" for its graphic descriptions and bloody photographs. Anslinger relied often in his gore file in the early years of the campaign, airing out lurid stories every time a Hearst newspaper[377] needed another salacious tale of violent miscegenation or the commissioner himself had to head down to the Capitol to beg for appropriations from a tight-fisted Congress.[378]

Over time, the gore file became ever more morbid and outrageous, descending unerringly into shameless race-baiting and prurient titillation. These were some of his "greatest hits":

> West Va.--Negro raped a girl eight years of age.
> Two Negros [sic] took a girl fourteen years old and kept

> her for two days in a hut under the influence of marihuana. Upon recovery she was found to be suffering from syphilis.
> Colored students at the Univ. of Minn. partying with female students (white) smoking and getting their sympathy with stories of racial persecution. Result pregnancy.
> Undercover agent invited to marihuana party. Suggestion that everyone take off their pants, both male and female. Agent dropped blackjack while disrobing and had to arrest immediately.
> Corpus Christi--Gov. of Texas told me of a case he knew about personally, and one which in some measure influenced him to destroy 600 acres of hemp. An oil worker, good character, smoked a cigarette, raped his six-year-old daughter. When his wife returned home in the evening, she found him lying across the bed in a stupor and the little child torn and bleeding. He couldn't remember. Was sentenced to death.

Yet despite the salacious spread of details evident in this list, Anslinger favored another story even more, the grisly tale of Victor Licata, a twenty-one year old man from Tampa, Florida, who murdered his entire family - his parents, two brothers, and sister - with an ax. The sensationalist news coverage the following day made bare mention of the marijuana connection:

> W.D. Bush, city detective chief, said he had made an investigation prior to the crime and learned the slayer had been addicted to smoking marihuana cigarettes for more than six months. This he said had unbalanced his mind at least temporarily.[379]

In truth, Licata had evidenced unbalanced behavior well before he ever smoked pot; the Tampa police had tried unsuccessfully to have the young man committed a full year before the vicious murders.[380] Conveniently ignoring such facts, Anslinger rested full responsibility at the feet of the monster marijuana, employing the story liberally in his journalistic propaganda campaigns.[381]

Anslinger did his job so well that public opinion began to accuse him of being too soft on crime. Caught between the rock of his family's petrochemical interests and the hard place of a terrified American public demanding federal prohibition, Anslinger leaned harder than ever on his crack legal team, Herman Oliphant and A.L. Tennyson, to find a way around the Constitution. Oliphant recommended the treaty approach - the same power that motivated Hamilton Wright and Stephen Porter to seek international conferences.[382] As it happened, the second meeting of the prestigious Geneva Conference was scheduled for later in 1936,[383] so Anslinger lobbied hard for the issue,[384] pushing the State Department to issue an ultimatum requiring some kind of anti-marihuana treaty as a requisite for U.S. support; but State vetoed the proposal. In any case, Oliphant realized that such pomp and circumstance was hardly needed for the Bureau's purposes; any treaty would do. So Oliphant cleverly designed a three-way treaty between Canada, the U.S., and Mexico which would require all three countries to pass federal legislation to deal with the marihuana menace.[385] State contacted Ottawa and found an enthusiastic partner there; but miscommunications with Mexico City delayed the passage of the treaty for months. Anslinger and his cronies at Treasury became impatient, demanding a solution which the lawyers couldn't see any way to solve. But then a dramatic Supreme Court decision suddenly pointed the way forward.

The National Firearms Act of 1934 represented the first successful attempt by Democrats to circumvent the Second Amendment by relying, oddly, on Congress' taxing power. Technically, the Firearms Act allowed anyone to purchase an automatic machine gun who wanted one; but anyone who chose to do so had to report the transaction on a government form and mail a copy to Washington with a check for $200, a "tax" on the transaction. Although the scheme's true intent was transparent, the Supreme Court nevertheless upheld the law on March 29th, 1937 by an emphatic unanimous opinion.[386] Oliphant suddenly knew exactly how to proceed; he quietly filed H.R. 6385, the Marihuana Tax Stamp Act of 1937,[387] two weeks after the Court's decision.

To prepare for the inevitable Congressional hearings, Anslinger began interviewing experts in the Surgeon General's office, in particular Dr. William Treadway, the head of the Mental Hygiene Division, for the task of intimidating Congress into action. But Anslinger didn't get the answers he wanted; to the question *What are the proofs that the use of marihuana in any of its forms is habit-forming or addictive, and what are the indications and positive proofs that such addiction develops socially undesirable characteristics in the user?* Treadway replied in full:

> Cannabis indica does not produce dependence as in opium addiction. In opium addiction there is a complete dependence when it is withdrawn there is actual physical pain which is not the case with cannabis. Alcohol more nearly produces the same effect as cannabis in that there is

> an excitement or a general feeling of lifting of personality, followed by a delirious stage, and subsequent narcosis. There is no dependence or increased tolerance as in opium addiction. As to the social or moral degradation associated with cannabis it probably belongs in the same category as alcohol. As with alcohol, it may be taken a relatively long time without social or emotional breakdown. Marijuana is habit-forming although not addicting in the same sense as alcohol might be with some people, or sugar, or coffee. Marijuana produces a delirium with a frenzy which might result in violence; but this is also true of alcohol.[388]

The response got Dr. Treadway disinvited from any further debate over the bill, so that the law-making apparatus of the federal government was forced to proceed without the advice of its most preeminent expert on the subject.[389] As a substitute, Anslinger found Dr. James Munch of Temple University, whose sole cannabis bona fides were a series of experiments on the effects of hashish extract on dogs.

In January of 1937, Anslinger convened a preliminary conference in the Treasury Building to hash out a few details. Of first importance was the question of what exactly to ban - the doctors and scientists present debated over what parts of the cannabis plant contained "active principles" and which could be allowed loopholes to allow the attendant industries to continue. Anslinger remained adamant: all forms of the cannabis plant had to go; allowing loopholes would only increase the cost of enforcement.[390] No one dared oppose him.

Recognizing the public relations difficulties of selling the American public on a new encroachment of federal power, a Treasury lawyer named S.G. Tipton quizzed Anslinger on whether he had any terrifying reports of the drug's effects: "horror stories - that's what we want."[391] Tipton needn't have worried; Anslinger's reliable "gore file" had grown thicker than ever. The group convened with the explicit goal of full prohibition.

And not a moment too soon. DuPont's investors had begun to protest to management, complaining that their radical petrochemical model wasn't making enough money.[392] The DuPont Board of Directors addressed the issue head-on in the company's 1937 annual report, reassuring their investors that the Board foresaw "radical changes" from "the revenue raising power of government … converted into an instrument for forcing acceptance of sudden new ideas of industrial and social reorganization."[393]

Within weeks of the annual report's release, with the party line well traced and undisturbed by an iota of medical evidence, Mr. Anslinger went to Congress.[394] For five morning sessions before the House Ways and Means Committee and one before the Senate Finance Committee,

Anslinger held his audience rapt with the "assassin of youth" routine, revealing one horror story after another. Not a shred of scientific evidence was presented; indeed, all of the government's most qualified experts, including Dr. Treadway and the architects of the Panama Canal Zone study, had disqualified themselves by telling the truth.[395] Instead, Anslinger appointed himself resident expert, expounding his cherished cannabis-crime theory and dismissing naysayers in white laboratory coats:

> Despite the fact that medical men and scientists have disagreed upon the properties of marihuana, and some are inclined to minimize the harmfulness of this drug, the records offer ample evidence that it has a disastrous effect upon many of its users. Recently we have received many reports showing crimes of violence committed by persons while under the influence of marihuana.[396]

To buttress his argument, Anslinger trotted out the gore file, horrifying the Senate Committee with gruesome photographs and horror stories of dubious veracity.[397] When an alarmed senator inquired how many joints would suffice to turn a user insane, Anslinger cinched the deal: "I believe in some cases one cigarette might develop a homicidal mania, probably to kill his brother. It depends on the physical characteristics of the individual."[398] At the House, Anslinger brought the amenable Dr. Munch, whose demonstrated acumen was so pathetic that lay Congressmen had to hold his hand through basic principles in order to put their desired responses on the record.[399] Yet if Munch were belittled, Anslinger's testimony was accepted unopposed, with all representatives unquestioningly recognizing his authority.

In fact, the only dissenting voice came from Dr. William Woodward, who appeared in a flustered state[400] the final day of the hearings. While continually reassuring the committee that the AMA had no objection to working with Congress to regulate cannabis, the clearly flabbergasted Woodward lambasted the thoroughly unscientific nature of the proceedings:

> That there is a certain amount of narcotic addiction of an objectionable character no one will deny. The newspapers have called attention to it so prominently that there must be some grounds for their statements. It has surprised me, however, that the facts on which these statements have been based have not been brought before this committee by competent primary evidence. We are referred to newspaper publications concerning the prevalence of

> marihuana addiction. We are told that the use of marihuana causes crime.
>
> But yet no one has been produced from the Bureau of Prisons to show the number of prisoners who have been found addicted to the marihuana habit. An informal inquiry shows that the Bureau of Prisons has no evidence on that point.
>
> You have been told that schoolchildren are great users of marihuana cigarettes. No one has been summoned from the Children's Bureau to show the nature and extent of the habit among children. Inquiry of the Children's Bureau shows that they have had no occasion to investigate it and know nothing particularly of it. Inquiry of the Office of Education - and they certainly should know something of the prevalence of the habit among the schoolchildren of the country, if there is a prevalent habit - indicates that they have had no occasion to investigate and know nothing of it.
>
> Moreover there is in the Treasury Department itself the Public Health Service, with its Division of Mental Hygiene [the Division headed by the skeptical Dr. Treadway]... No one has been summoned from that Bureau to give evidence on that point. Informal inquiry by me indicates that they have had no record of any marihuana or cannabis addicts who have even been committed to [government treatment farms at Lexington, Kentucky].
>
> The Bureau of the Public Health Service has also a division of pharmacology. If you desire evidence as to the pharmacology of cannabis, that is obviously the place where you can get direct and primary evidence, rather than the indirect hearsay evidence.[401]

Woodward also referred to his work in drafting the Uniform Narcotic Acts, during which he discovered that most of the rumors of cannabis's dangers and reports that its use among the youth had increased were nothing more than myths. Finally, he noted that effective federal enforcement of such a law would be practically impossible anyway, since the plant grew so abundantly as to make its complete elimination a Herculean task requiring massive appropriations. Darkly, he noted the certain and insidious squelching of medical research into cannabis' further benefits - a wry reference to what he and others suspected may have been Anslinger's true intent.[402]

As soon as the doctor had finished his preliminary statement, each member of the committee took turns excoriating him. Congressman Vinson accused Woodward and his association of superfluous obstructionism; Congressman Cooper accused him of evading the issue; Committee Chairman Doughton accused the AMA of petulantly objecting to the bill just because they weren't consulted in its drafting. After subjecting Woodward to a legislative gauntlet as hostile as Anslinger's reception had been deferential, the committee brusquely dismissed Dr. Woodward without even thanking him for his testimony.[403] Both committees recommended the bill's passage over the AMA's objection, and Chairman Doughton took it to the House floor late in the day of June 10th, after a long and sweltering day of debate. When Congressman Snell objected to the bill's sudden springing at the late hour with virtually no discussion, one of his colleagues stepped in:

> Mr. RAYBURN: Mr. Speaker, if the gentleman will yield, I may say that the gentleman from North Carolina has stated to me that this bill has a unanimous report from the committee and that there is no controversy about it.
> Mr. SNELL: What is the bill?
> Mr. RAYBURN: It has something to do with something that is called marihuana. I believe it is a narcotic of some kind.
> Mr. VINSON: Marihuana is the same as hashish.
> Mr. SNELL: Mr. Speaker, I am not going to object but I think it is wrong to consider legislation of this character at this time of night.[404]

After other Representatives insisted on discussing the matter, Congressman Vinson - who had hostilely attacked Dr. Woodward at the Ways and Means Committee hearings - silenced debate by confidently stating that the AMA representative (a "Dr. Wharton") had fully endorsed the bill; dishonestly reporting unanimous support. When the Speaker brought the bill to a vote, he declared the bill's passage without even calling the roll.[405] The Tax Stamp Act became law on August 2nd, 1937, with the only coverage by the New York Times reading in its entirety: "President Roosevelt signed today a bill to curb traffic in the narcotic, marihuana, through heavy taxes on transactions." The revenue-raising arm of government had forced radical new ideas of social reorganization, just as DuPont had predicted.[406]

The effect was immediate. While street dealers easily evaded the FBN's 250 agents, the otherwise law-abiding farmers who made no attempt to hide were forced through a slow death by bureaucracy.[407] The listing for cannabis disappeared from the US *Pharmacopeia*, victim of covert pressure from Anslinger.[408] Henry Ford, stymied by the mountain of paperwork and punitive fines, had to shelve his ideas for a sustainable car; hemp farmers of Wisconsin and Kentucky, inheritors of a proud tradition, buckled under the oversight and closed up shop. The land at Arlington Farms, where Lyster Dewey bred the strains to rescue his country, was paved over with asphalt to become one of the parking lots of the Pentagon.[409] With astonishing speed the American hemp industry died out altogether, the last Wisconsin hemp farmer begrudgingly plowing his last hemp field over, saving the last few sacks of precious seed to deliver to a U.S. government suppository. Yet the government caretakers never even put the seeds in storage, casting them instead into a dark, lonely corner - where the dream of a renewed America, independent of powerful, selfish interests,[410] went extinct.[411]

EPILOGUE

If anyone doubted whether Anslinger would aggressively enforce the new law, the determined bureaucrat put all such doubts to rest. Anslinger personally attended the first federal trial of an accused Tax Stamp Act violator, a 58-year old Denver man named Samuel Caldwell.[412] After the judge sentenced Caldwell to a $1,000 fine and four years' hard labor in Leavenworth, Anslinger triumphantly proclaimed that the federal government would enforce the new law to the letter.[413]

He retreated almost immediately. The early years of World War II posed a thorny roadblock to Anslinger's goal of full cannabis prohibition, when in the moment of one of the nation's greatest crises the United States suddenly remembered hemp's importance. After the Philippines, the source of the majority of the rope and twine used by the U.S. military, fell to the Japanese in 1942, the U.S. Navy - which played a pivotal role in supplying the Allied powers in Europe - found itself subject to a monumental need for strong, durable fibers, and nowhere to obtain them.[414]

Until they remembered their history. The remnants of the U.S. industrial hemp base, worn to tatters by the FBN, suddenly exploded.[415] At the request of the Department of Defense, 36,000 acres of hemp were sown in 1942, a number that grew to 50,000 the following year. With the tons of hemp fiber yielded through the program, the Defense department made canvas tarpaulins, soldier's rucksacks, parachutes - but most importantly rope for the legendary Liberty ships, the fleet of supply vessels which overwhelmed the German U-boat blockade with sheer numbers, bringing the supplies the Allied forces needed to turn the tide.[416] But while the needs of war materiél checked the FBN temporarily, the more lasting threat to Anslinger's plan had come from Anslinger himself.

The manufactured menace which he created had become a monster unto itself, more fearsome, perhaps, than even the media-savvy bureaucrat could handle. With public fear already at a fever pitch, every new marijuana arrest - especially of a white youth "from a good family" - triggered knee-jerk bouts of hysteria so severe that the phone lines at Narcotics clogged with frenzy. Even more damaging was a legal theory beginning to take hold among defense lawyers, a logical extension of the FBN line.

The idea of the marijuana insanity defense plea originally came, like so much else of America's cannabis history, from Mexico. In the wake of 19th century state prohibitions of *mariguana* on the theory that its use caused violent fits of fury, defense lawyers began to argue quite for the policy's exculpatory implications. After all, if it were proven that the defendant had smoked a joint before committing murder, he couldn't really be held responsible for his crime, if all that was said of *mariguana* were true.[417] In a number of well-publicized cases, defense lawyers in the U.S. attempted to adapt the argument for the benefit of their accused clients; [418] and although none ever succeeded in winning a not guilty verdict on the basis of the defense, the publicity nevertheless embarrassed Anslinger's Bureau. One of the most embarrassing moments came courtesy of none other than Dr. James Munch, the Temple University cannabis "expert" who had been brought in to render medical opinions on marijuana after the more eminent government experts had disqualified themselves by speaking truth to power. When called to offer expert testimony in the case of an accused murderer who had been a regular user of "marijuana cigarettes," Munch held the jury in thrall with his report of taking a trial with grass himself, in which he found himself trapped at the bottom of an ink bottle for 200 years.[419] The law and order brigade was scandalized when Munch's testimony let the killer off easy with life imprisonment instead of the electric chair.

Anslinger managed to muzzle Munch,[420] but the much wider problem remained. The bureaucrat had gone too far and held too much at stake to reverse his course - yet some way to deflect the hype of the marijuana menace had to be found. Instead of an insidious enemy creeping around every corner and lurking beneath every window, the savvy PR expert realized he needed an extrinsic enemy. An external threat. Some sort of group - indelibly associated with cannabis - which the majority of America could readily fear without having to understand, could demonize without hope of escape. By playing intelligently on old and deep-seated rivalries, the crafty, tortured man hoped to attach the specter of the marijuana menace to a sustainable fixture of American society. It only took the appearance of the right group - alien, misunderstood, and frightening.

In fact, the seeds of such a group had been sown long before. Having already taken root, the group was growing quickly, right under his nose.

About the Author

Jeremy Daw began his odyssey with cannabis as a graduate student at New York University in 2002. At the age of 22, debilitating nausea had become a daily fact of life, and his weight was dropping to dangerous levels. After no either treatment proved effective, he turned to cannabis, which immediately restored his appetite. He enrolled at Harvard Law School's *juris doctor* program in the fall of 2004, where he immersed himself in the study of US drug policy and its constitutional limits. Since graduating law school, he has written extensively about the challenges of constantly changing laws. He is Editor-in-Chief of *Cannabis Now Magazine* and a contributor to the *West Coast Leaf.* He lives and writes in sunny Berkeley, California.

Follow Jeremy on:

Twitter: @weed_the_people
Facebook.com/JeremyDawJD
www.weedthepeoplebook.com

FURTHER READING

Abel, Ernest. *Marihuana: The First Twelve Thousand Years*. New York: Plenum Press, 1980.

Bonnie, Richard and Charles Whitebread. *The Marijuana Conviction: A History of Marijuana Prohibition in the United States*. New York: The Lindesmith Center, 1999.

Booth, Martin. *Cannabis: A History*. New York: St. Martin's Press, 2003.

Bourrie, Mark. *Hemp: A Short History of the Most Misunderstood Plant and its Uses and Abuses*. Buffalo, New York: Firefly Books, 2003.

Campos, Isaac. *Home Grown: Marijuana and the Origins of Mexico's War on Drugs*. Chapel Hill: The University of North Carolina Press, 2012.

Caulkins, Jonathan, Angela Hawken, Beau Kilmer and Mark Kleiman. *Marijuana Legalization: What Everyone Needs to Know*. Oxford University Press, 2012.

DeQuincey, Thomas. *The Confessions of an English Opium Eater*. London: Bodley Head, 1930.

Dewey, Lyster H. and Jason L. Merrill, "Hemp Hurds As Paper-Making Material," Washington, D.C.: USDA Bulletin No. 404, October 14th, 1916

Dietch, Robert. *Hemp: American History Revisited, The Plant with a Divided History*. Algora Publishing, 2003

Dulchinos, Donald. *Pioneer of Inner Space: The Life of Fitz Hugh Ludlow, Hasheesh Eater*. New York: Autonomedia, 1998.

Fortenbery, T. Randall and Michael Bennett. "Is Industrial Hemp Worth Further Study in the U.S.? A Survey of the Literature" *Agricultural and Applied Economics* Staff Paper No. 443, 2001.

Grinspoon, Lester. *Marihuana Reconsidered*. Cambridge: Harvard University Press, 1971.

Grinspoon, Lester and James B. Bakalar. *Marihuana, The Forbidden Medicine*.

New Haven: Yale University Press, 1993.

Grotenhermen, Franjo and Ethan Russo, eds. *Cannabis and Cannabinoids: Pharmacology, Toxicology, and Therapeutic Potential.* New York: The Haworth Integrative Healing Press, 2002.

Herer, Jack. *The Emperor Wears No Clothes.* Austin, Texas: AH HA Publishing Co, 2000.

Holland, Julie, Ed. *The Pot Book: A Complete Guide to Cannabis, Its Role in Medicine, Politics, Science and Culture.* Rochester, Vermont: Park Street Press, 2010.

Hopkins, James. *A History of the Hemp Industry in Kentucky.* University Press of Kentucky, 1998.

Industrial Hemp Investigative and Advisory Task Force Report. Submitted to the Illinois House of Representatives January 26, 2000.

Jonnes, Jill. *Hep-Cats, Narcs, and Pipe Dreams: A History of America's Romance with Illegal Drugs.* Baltimore: The Johns Hopkins University Press, 1996.

Latimer, Dean and Jeff Goldberg. *Flowers in the Blood: the Story of Opium.* New York: F. Watts Co., 1981.

Lee, Martin. *Smoke Signals: A Social History of Marijuana - Medical, Recreational, and Scientific.* New York: Scribner, 2012.

Lindop, Grevel. *The Opium-Eater: A Life of Thomas DeQuincey.* New York: Taplinger Publishing Company, 1981.

Kraenzel, David G., Tim Petry, Bill Nelson, Marshall J. Anderson, Dustin Mathern and Robert Todd. "Industrial Hemp as an Alternative Crop in North Dakota: A White Paper Study of the Markets, Profitability, Processing, Agronomics and History." Agricultural Economics Report no. 402. North Dakota State University: Institute of Natural Resources and Economic Development, 1998.

Ludlow, Fitz Hugh, *The Hasheesh Eater* New York: Harper Brothers, 1857.

Mikuriya, Tod, ed., *Marijuana: Medical Papers, 1839-1972* (Oakland: Medi-Comp, 1973).

Mills, James. *Cannabis Britannica: Empire, Trade and Prohibition 1800-1928.* Oxford University Press, 2005.

Musto, David. *The American Disease: Origins of Narcotic Control*, 3rd ed. Oxford University Press, 1999.

Ranalli, Paolo and Gianpietro Venturi. "Hemp as a raw material for industrial applications," *Euphytica* 140, 1-6. 2004.

Rosevear, J., *Pot: A Handbook of Marihuana* New Hyde Park, N.Y., 1967.

Schultes, Hofmann, and Rätsch. *Plants of the Gods.* New York: McGraw-Hill, 1979.

Sloman, Larry. *Reefer Madness: A History of Marijuana.* New York: St. Martin's Griffin, 1979.

Small, Ernest and David Marcus. "Hemp: A New Crop with New Uses" in *Trends in New Crops and New Uses.* Alexandria, Virginia: ASHS Press, 2002.

Werner, Clint. *Marijuana: Gateway to Health.* San Francisco: Dachstar Press, 2011.

NOTES

[1] See Booth, p. 2.
[2] Booth, p. 20; Caulkins et al, p. 18.
[3] Booth, p. 20.
[4] Booth, p. 22.
[5] Booth, p. 24.
[6] See Abel, p. 76.
[7] "There is probably not a country in the world, save those with very cold climates, where cannabis is not to be found growing either wild or in cultivation. It exists from Manchuria and the steppes of Mongolia to the rural shires of southern Britain, from the Ganges plain in India to the prairies of the American Mid-West and Hokkaido in Japan. Under cultivation, it is to be found throughout the Indian sub-continent, South-East Asia, the Caribbean, Central and South America, most of sub-Saharan and parts of north Africa, and southern Europe - and that is just in the open, not under glass. It is, without a doubt, the most widely distributed hallucinogenic plant on the planet" (Booth, pp. 17-18).
[8] Technically, cannabis "seeds" are not actually seeds but *achenes*, a form of fruit (Sloman, p. 53).
[9] Dietch, p. 30.
[10] This story comes from the Boston Massacre Historical Society website, www.bostonmassacre.net.
[11] Accounts vary. See, e.g., *A Short Narrative*, or Ferling, John. *John Adams: A Life* (1992).
[12] From Adams, *A Short Narrative.* See also Bourrie, pp. 18-20.
[13] Quotations from the transcript of the trial of *Rex v. Weems et al*, the case against the soldiers. John Adams for the defense.
[14] Booth (2004), p. 35, Dietch p. 21. According to Herer, p. 7: "Until 1883, 75-90% of all the paper the world produced was made with hemp fiber." Bibles in particular have traditionally been printed on paper made from hemp, because the thinness of their leaves requires a material stronger than wood pulp to prevent tearing (Grinspoon, pp. 10, 13). In truth, hemp paper goes back much further than Gutenberg; according to Abel (pp. 6-7), the very first true paper in history, invented by the Chinese, was made from mulberry bark and hemp fiber.
[15] Dietch, p. 35. Herer (Chapter 2) claims that "our forbears were well aware that hemp is softer than cotton, warmer than cotton… Were it not for the historically forgotten (or censored) and currently disparaged marijuana plant, the Continental Army would have frozen to death at Valley Forge, Pennsylvania."
[16] See Kraeznel (1998); Illinois Industrial Hemp Investigative and Advisory Task Force Report (2000). Dietch, p. 35: "Betsy Ross purportedly made the first flag from hemp cloth and until 1937 all American flags were made from hemp." If the first American flag was made of hemp fiber, such would be in keeping with a long tradition of making flags out of hemp to come out of England (Grinspoon p. 11).
[17] The "L" is for Carl Linnaeus, who in 1753 gave cannabis its binomial scientific name still in use today (Campos p. 68; Grinspoon, p. 1; Bonnie and Whitebread, p. 1). "Sativa" means "cultivated" in Latin.

[18] "Cultivation of the plant for its fibrous content was practically simultaneous with the founding of the early American colonies… The climate of the American colonies was considered ideal for growing hemp and flax, and the English looked forward to an abundant yield. King James I… firmly corrected Sir Walter Raleigh when the latter suggested that energy might be better put to the cultivation of tobacco, which was just being introduced to Europe. The Virginia Company made the cultivation of hemp a part of its contract agreement." (Grinspoon, pp. 10-11).
[19] Campos, p. 1. Booth (p. 38) notes that Quadrado was "a *conquistador* serving with Cortés."
[20] This was probably the first introduction of cannabis in the New World (Campos pp. 1, 39), but see Booth (pp. 38-39) reviewing theories that it had been brought by Vikings, Chinese sailors, or even across the Bering land bridge during the Ice Age.
[21] Hemp's companion fiber, flax, has grown beside it in European-controlled farms seemingly since hemp first arrived. Booth (pp. 34-35) notes "[t]he first recorded growing of hemp [in England] - with flax - was at Old Buckenham, fifteen miles south-west of Norwich. By 600 [C.E.], with the arrival of the Saxons, hemp growing was well established." Alexander Hamilton wrote as Treasury Secretary that "Flax and Hemp: Manufacturers of these articles have so much affinity to each other, and they are so often blended, that they may with advantage be considered in conjunction" (Herer, Chapter 2).
[22] *Recopilación de leyes de los reynos de las Indias* (1973), p. 117. Quoted in Campos, p. 52. Bonnie and Whitebread (pp. 2-3) attribute Almagro and Valdivia as the first Spaniards to sow hemp in Chile.
[23] Dietch (p. 13): "The British Crown mounted the Roanoke expedition (1582; it failed) and established the Jamestown Colony (1607) in an effort to explore possibilities in America… primarily because Britain's domestic hemp-based industry, the lifeblood of the economy, desperately needed a stable, reliable, and relatively cheap source of raw hemp." Campos (p. 40): "Cannabis fiber had thus literally become the stitching of maritime empire." See also Hopkins, *History of the Hemp Industry*; and Mills, *Cannabis Britannica*, pp. 17-18.
[24] Booth (p. 37), says 80 tons; Campos (p. 40) says 90. Herer (Chapter 2) claims that "it cost more for a ship's hempen sails, ropes etc. than it did to build the wooden parts."
[25] See Abel, p. 53.
[26] For descriptions of the retting process, see Booth, p. 43; Bourrie, pp. 13-14; Fortenbery and Bennett, p. 10.
[27] Booth, p. 37.
[28] Abel, p. 46; Booth, p. 37; Dietch, p. 12; Grinspoon, p. 11; Bourrie, p. 14. In 1533 Henry levied a fine of 3s. 4d. on every 60 acres which failed to devote at least a quarter acre to hemp.
[29] Booth, p. 37.
[30] Zinn, Howard. *England and the Baltic in the Elizabethan Era*, p. 232. Grinspoon (p. 11) attributes the balance of England's hemp imports to the Dutch East Indies; a trade war shortly before the founding of Jamestown intensified England's need for an independent source of hemp. See also Fortenbery and Bennett, p. 6; Abel, p. 46.
[31] Dietch, p. 14; Abel, pp. 40-43.

---

[32] Dietch, p. 14; Abel, p. 47.
[33] Booth, pp. 37-38: "The defeat of the Spanish *Armada* established England as the primary European maritime power. Consequently, the need for hemp increased further and, even had the English farmers been willing, they would have been hard-pressed to meet the demand. Importation was essential with English vessels visiting Danzig, Riga, and St Petersburg for their cargoes… Such reliance upon a foreign producer not only all but killed off the English hemp trade but, as the Venetians discovered, it also left the country vulnerable. Britain had to turn to its fledgling colonies in the New World for supply."
[34] Booth, p. 38.
[35] Cortés famously wrote back to the Crown that there would be no need to send physicians to the New World because the indigenous of Mexico were already so advanced in this regard. Aguirre Beltrán, *Medicina y magia*, p. 80.
[36] For an excellent review of Pre-Columbian indigenous medical practices, see Campos, pp. 41-44.
[37] Schultes, Hofmann, and Rätsch, p. 26; Díaz, "Ethnopharmacology of Sacred Psychoactive Plants," pp. 654-55.
[38] José de Acosta, *Historia natural y moral de las Indias* (1590).
[39] Quoted in Campos, p. 47.
[40] Not the first anti-drug law on Earth. In an eerie demonstration of history repeating itself, "[i]n 1231, Pope Gregory IX initiated the Holy Inquisition, its aim to root out and destroy the heresies of the previous two centuries. Hemp, being regarded as sorcerous, was outlawed as heretical. Those who used it, whether for medical purposes or divination, were branded as witches who, in turn, were considered heretics.

The persecution of witches across Europe commenced in earnest in 1484… in order to prevent the celebration of the Black or Satanic Mass, Innocent VIII banned the use of hemp in ritual… cannabis was said to be a vital ingredient of witches' brews. Used to formulate the 'black sacrament' with opium, hemlock, belladonna and other magical or poisonous plants, it supposedly aided in driving the Satanists into an ecstatic frenzy, making them hungry and acting as an aphrodisiac to ready them for their orgies. When not employed ritually, hemp seed oil was purportedly a major constituent of 'flying ointment', that which witches used to 'ride their broomsticks'" (Booth pp. 71-72). It was under this period of Catholic prohibition that Francois Rabelais (1483-1553) wrote his satirical Gargantua and Pantagruel, in which "the herb pantagruelion" is a thinly-veiled reference to hemp. The risk Rabelais took was not minor; when the Inquisition discovered that the famed Portuguese doctor and botanist Garcia Da Orta was secretly Jewish, they nearly succeeded in destroying every copy of his life's work (Booth p. 74).
[41] Booth (p. 36) notes that the sails used in the late 16th century to maneuver in open ocean "need[ed] to be very durable and tear resistant, for [they were] subjected to considerable strain. Hemp gave… the necessary qualities."
[42] See Abel, p. 46.

[43] Booth pp. 39-40: "Settlers at Jamestown preferred to grow more lucrative tobacco but cultivated hemp under a contract signed with the Virginia Company in 1607, just prior to the founding of the colony."
[44] Booth, p. 39.
[45] Booth, p. 40; Herer, Chapter 1.
[46] Booth, p. 40.
[47] Grinspoon, p. 11; Booth, p. 40.
[48] Booth, p. 40: "When the tobacco market slumped, as it did periodically, hemp kept the farmers' cash flow going."
[49] Abel, p. 50. Booth, p. 40: the colonial government set aside a budget of "ten guineas per man" to induce the migration of skilled hemp dressers.
[50] Booth, p. 33.
[51] Dietch, p. 17.
[52] See Booth, p. 34; Dietch, pp. 20-21.
[53] Rosevear, pp. 20-21; Grinspoon, p. 11; Booth, p. 40.
[54] This and the laws which follow can be found listed in Booth, pp. 40-41.
[55] Booth, p. 40. Bourrie (p. 16) claims that John Harrison's ropewalk was "one of the city's most important employers and a focus of discontent that would contribute to the local desire for independence."
[56] Sloman, p. 21; Booth, p. 40.
[57] Abel, p. 50; Deitch, p. 17.
[58] Hemp proved valuable to the colonies in more ways than one. See Booth, pp. 40-41: "[H]emp production rose in Maryland and Virginia but little of the harvest reached Britain… In the decades following the end of the war, hemp became more than a commodity: it became a currency… Because of the lack of confidence in local currencies and paper money, which was readily counterfeited, hemp became a currency in itself. Hemp's more or less constant quality and the permanent demand for it made it ideal for barter." See also Abel pp. 57-58; Herer, Chapter 1.
[59] Abel (p. 55) notes that the arrival of the professional European spinners elicited excited responses from the women of the city, who flocked to their factories to learn how to spin better thread. Booth, p. 33. After a disastrous fire in Boston in 1794, the city's ropewalks were encouraged to relocate to the newly-drained Back Bay, which now is one of the city's trendiest and most expensive residential neighborhoods (Bourrie p. 17).
[60] Deitch p. 19, 25-7, Booth p. 34-5; see also Fortenbery and Bennett, p. 6.
[61] Deitch (2003), p. 19; Herer, Chapter 1.
[62] According to Deitch (p. 21) and Herer, (p. 7), until 1883, 75-90% of all paper produced worldwide was made of cannabis fiber.
[63] See Dewey and Merrill, 1916.
[64] Dewey concluded that 1 acre of cannabis could sustainably produce an amount of paper equivalent to 4.1 acres of forest land. *Ibid.*
[65] Dietch, p. 19; Booth, p. 43; Bourrie, p. 21.
[66] Indeed, by 1820, Jefferson had given up cannabis production entirely, being only partly satisfied with the performance of his invention. He wrote that "the beating and breaking [of] it, which always has been done by hand, is so slow, so laborious,

and so much complained of by our laborers, that I have given it up." See Booth, p. 42.
[67] Deitch (2003), p. 29
[68] From "Hemp for Victory!" Short film by the U.S. Dept. of Agriculture, 1942.
[69] Booth, p. 42.
[70] Indeed, by the time of the Revolutionary War the Colonies had produced such a surplus that they began to trade with European countries other than England: e.g. France (Booth, p. 41).
[71] The practice of naming cannabis strains after their country of origin continues to the present day, indicated by strain names such as "Maui Wowie," "Acapulco Gold," and the vast array of "Kushes," eponymous with the land of their ancestors, the Hindu Kush Mountains between Afghanistan and Pakistan. Today, many pure indica and indica-heavy blends are more commonly prescribed for patients with severe chronic ailments, because of the indica tendency to favor a stronger balance of those cannabanoids which have the greatest medical benefit but not the strongest "high", unlike THC.
[72] Deitch, p. 26. But see Sloman, p. 21.
[73] Deitch, p. 30.
[74] Deitch, p. 25. For more quotes from Washington's agricultural diary at Mount Vernon, see Sloman, p. 21.
[75] Grinspoon p. 12 (citing E.M. Houghton and H.C. Hamilton, "A Pharmacological Study of Cannabis Americana," *Amer. J. Pharmacol.,* 80 [1908], 17) disputes that fertilized female cannabis flowers are less psychoactive than those which are unfertilized. Even so, it does not prove that common wisdom didn't believe this to be the case in Washington's time (as it is today).
[76] Deitch, p. 25.
[77] Deitch, pp. 25-26.
[78] Deitch, p. 26.
[79] Abel, p. 55; Booth, p. 41.
[80] "By 1770, the finances of the East India Company… were in a shambles due to a famine in Bengal. The board was forced to approach Parliament for a one-million-pound loan to prevent its collapse" (Booth p. 75). Abel (p. 78) blames the company's bankruptcy on embezzling by top company executives.
[81] Booth, p. 34. In part, the surpluses may have resulted from colonial laws exempting from military service any farmer who devoted at least six months to producing cannabis rope and cloth for the war effort (Abel, p. 56).
[82] Booth, p. 34.
[83] Booth, p. 41: "By the outbreak of the War of Independence, the colonies were more or less self-sufficient in cloth - and hemp. Any surplus that arose was exported - mostly to the French, Britain's main mercantile competitor and enemy of long standing - with the proceeds used to purchase munitions."
[84] In "Common Sense," Thomas Paine lists the more resources most necessary to fight for independence: "cordage, iron, timber and tar." Regarding the first, "[h]emp flourishes even to rankness, we do not want for cordage."

---

[85] Booth (p. 43) notes that although hemp production declined for most of the states in the early 19th century, hemp persevered as the primary cash crop of the bluegrass region of Kentucky from 1810 to 1889.
[86] Fortenbery and Bennett, p. 6: "Reasons [for the decline of the US hemp industry] include the development of the cotton gin (which reduced labor costs for Southern cotton production)."
[87] "Hemp accounted for 5 percent of the weight of a cotton bale, and the fortunes of the Kentucky industry rose and fell with the cotton market" (Bourrie, p. 21). One of the greatest benefits of a continuing hemp industry, from a humanitarian point of view, is the result of the customs of the hemp production industry. As Booth (p. 44) notes, "[d]espite the hard work, the slaves preferred hemp growing to other tasks. They were usually set a daily target which meant they had time off if they reached it early. Furthermore, some plantation owners paid 'wages' to encourage production. At harvest time, some slaves were given one cent per pound of fibre [sic] broken in a day, over a minimum of 100 pounds. With much sweating, a sturdy young male slave could earn up to two dollars a day. Some earned enough over a few seasons to buy their liberty." See also Abel, pp. 59-60. This fact, of course, arises not from any intrinsic botanical property of the plant itself but rather the economics of the market surrounding it - which also had darker implications: "Slaves who understood the cultivation and preparation of hemp, or worked in rope-walks, were amongst the most sought after and therefore expensive."
[88] The name "marijuana" comes from *mariguana*, a Christianization of *pipilzintzintlis* during the period of the Inquisition. Cannabis was also called *Rosa maria*, although the practice has mostly died out (Campos pp. 60-61, 74-78).
[89] Between 1824 and 1838, four separate research institutions had been founded in Mexico to study indigenous flora (Campos pp. 71-72).
[90] Campos, pp. 69-70.
[91] Campos, pp. 124-132. White Europeans and Americans did not extend the concept of racial degeneracy to Latinos only. "The American explorer Henry M. Stanley, who considered Africans to be a sub-species of humanity, said cannabis weakened them to the point that they were useless as expedition porters." Booth, p. 56.
[92] Campos, pp. 72-73.
[93] To cite only one pioneering example. Campos (pp. 182-84) gives credit to the more evolved medical theories of Muslim culture in the Middle Ages, when much of Spain had been under Muslim rule for centuries.
[94] See Campos, pp. 184-86.
[95] Campos, p. 193.
[96] American settlers headed west, as a rule, used hemp products for a variety of applications. "Apart from the harness ropes and canvas awnings of the 'prairie schooner' covered wagons, [hemp] was an ideal pioneer crop providing both a basic fuel and fibre [sic]. Blankets, tents, cord and tough work clothing were all made out of it although not, as many believe, jeans which were made out of duck, a heavy cotton fabric" (Booth, p. 45). See also Grinspoon, p. 10.
[97] Campos, p. 116.

[98] Campos, pp. 186-87. For a summary history of the *herbolarias* in Mexico, see *ibid* pp. 145-150.
[99] Some believed that Mexico's need to abstain from cannabis was even greater than elsewhere in the world. "The use of Indian hemp is found in all the regions where the plant grows and it is consumed as a solid or liquid, or as smoke, moreover I believe in no part of the globe does it produce such fatal effects as in the Mexican Republic" José Olvera, *Expendio libre de yerbas medicinales*, pp. 268-70.
[100] For more information on the variability of cannabinoid production in different climates, see Small, *Species Problem in Cannabis*, 1:79; Fetterman et al., "Mississippi-Grown Cannabis sativa L."; and Hemphill, Turner, and Mahlberg, "Cannabinoid Content."
[101] The word means "the most noble princes" (Campos, p. 1).
[102] Abel (p. 64), Bonnie and Whitebread (pp. 4-5) and Sloman (p. 29) hypothesize that the first Americans to smoke cannabis in a relaxational sense may have been African slaves working hemp fields. By the time of the founding of the New World colonies, the custom of smoking cannabis had already pervaded many far-ranging regions of Africa. From Booth, pp. 53-54: "Under Arabic influence [Arabic culture had adopted Hindu customs of smoking hashish at least as early as the 11th century], cannabis use spread across North Africa and south into sub-Saharan eastern Africa, although even an approximate date for this expansion outside the Islamic sphere is uncertain… Ethiopia was trading with the Arabs since at least the second century…

Wherever Arab traders went in Africa, cannabis was introduced. From about the third century onwards, they were a frequent presence along the East African coast where, by the twelfth century, there were Muslim settlements on the islands of Zanzibar, Pemba and Kilwa, with mainland Arab centres [sic] at Lamu, Gede and Sofala, south of Beira in Mozambique, to name but a few. They also traveled up navigable rivers, such as the Zambesi, searching for slaves. In this quest, they reached at least as far as eastern Zambia… There is even a suggestion that cannabis might have reached East Africa carried by the Chinese: Chinese traders were certainly visiting the coastal areas of Kenya as early as the late thirteenth century.

Arab incursion into central Africa took cannabis westwards whilst Africans themselves extended its range southwards to the Hottentots, Bushmen and 'kaffir' tribes of South Africa such as the Xhosa and Zulu peoples. In southern Africa it became known as *dagga* whilst in western Africa it was called *diamba*: today *dagga* is widely used across Africa, the name sometimes applied to any psycho-active plant material.

By the time Europeans started to reach eastern Africa, cannabis use was well established, both as a recreational drug and in a religious context."
[103] The British imported Indian indentured servants in the mid-1800s to almost anywhere in their empire, including South Africa, where they discovered to their dismay that the imports brought the Indian custom of growing cannabis with them (Booth pp. 140-41).
[104] "By 1820, interest in the exotic Orient was widespread. Chinese porcelain, silks and brocades were the height of fashion, Persian carpets adorned the floors and Indian tapestries the walls of the great houses of Europe, whilst Chinese and

---

Japanese craftsmen were manufacturing items specifically for Western markets" (Booth p. 79).
[105] As late as 1900, popular Chinese sentiment still held opium responsible for "sapping the strength and initiative of the nation." Musto p. 29.
[106] Campos, p. 51.
[107] Campos, pp. 132-38.
[108] Campos, p. 142.
[109] Coleridge was hardly the only of his contemporaries to know the vagaries of opium. "Some of the greatest writers in European literature fell under opium's spell - Wordsworth, Pushkin, Coleridge, Shelley, Goethe, Keats, Byron and Scott, to name but a few. Some were to pay the heavy price of opiate addiction and come to wonder if there might not be another drug which was not so cruelly habit-forming" (Booth p. 80). Not long later, across the English Channel, a group of French writers calling themselves "Le Club des Hachichins" regularly used cannabis extracts with the specific purpose of tapping into new wells of imagination. The club included Pierre Jules Théophile Gautier, Charles Baudelaire, the painter Fernand Boissard de Boisdenier, Gérard de Nerval, Victor Hugo, Honoré de Balzac, and the painter Honoré-Victorin Daumier as regular members; the painter Eugene Delacroix and Alexandre Dumas are rumored to have stopped by from time to time (Booth pp. 82-84).
[110] "Here for the first time the Western world was presented with an extraordinary new concept: drug addiction" (Jonnes, p. 15). DeQuincey was "the prodigal son of a wealthy Manchester mercantile family who ran away from home, lived in poverty in London with a prostitute then, accepted back by his family, went up to Worcester College, Oxford with the intention of becoming, as he would have it, 'the intellectual benefactor of mankind.' Whilst a student, he was prescribed laudanum (opium dissolved in fortified wine) for a bout of neuralgia and became addicted. In 1822, at the age of thirty-six, DeQuincey published his autobiographical *Confessions of an English Opium Eater* which had appeared serialized in the *London Magazine* the year before. The book immediately drew considerable attention, going into detail about the effects of opium on the imagination and the considerable - and terrible - consequences of addiction. For the first time, the effects and dangers of narcotics were broadcast widely, to become a major matter for discussion at every level of society" (Booth, p. 81).
[111] See Sonnedecker, Glenn, "Emergence of the Concept of Opiate Addiction," *J. Mondial Pharmacie*, Sept.-Dec. 1962, pp. 275-90; Jan.-Mar. 1963, pp. 27-34. D.I. Macht, "The History of Opium and Some of Its Preparations and Alkaloids," *JAMA* 64: 477-81 (1915). Terry, Charles and Mildred Pellens, *The Opium Problem* (New York: Bureau of Social Hygiene, 1928; reprint ed., Montclair, N.J.: Patterson Smith, 1970).
[112] Musto (3rd ed., 1999), p. 1, ft. 2 notes that at least one "extensive report" on the causes of the 19th century opiate epidemic (F. E. Oliver, "The Use and Abuse of Opium," Third Report of the Massachusetts Board of Health, 1871, pp. 162-77) placed more blame on the effects of Temperance campaigns in the 1840s and 1850s than on the effects of the Civil War.

[113] Musto (p. 13) notes that "if the physician could not effect cure, he could assuage pain and apprehension: opiates were preeminent for the functions and were apparently used with great frequency." DeQuincey himself "had begun using opium to alleviate an ailment, in this case rheumatic pains that struck in 1804 when he was a student at Oxford" (Jonnes, p. 15).
[114] Norman Howard-Jones, "A Critical Study of the Origins and Early Development of Hypodermic Medication," *J. Hist. Med.* 2:201-49 (1947). It is interesting to note that the Confederacy, in an attempt to thwart the Union blockade, made trials of growing opium poppies to augment their morphine supplies. Norman H. Frank, *Pharmaceutical Conditions and Drug Supply in the Confederacy* (Madison, Wis.: Institute of the History of Pharmacy, 1955). See also Bourrie, p. 40.
[115] See Bonnie and Whitebread, p. 8.
[116] Jonnes, p. 16. In 1680, Thomas Sydenham declared, "Among the remedies which it has pleased Almighty God to give man to relieve his sufferings, none is so universal and efficacious as opium."
[117] Musto p. 6: "Along with his prejudice [against Chinese immigrants] came a fear of opium smoking as one of the ways in which the Chinese were supposed to undermine American society."
[118] See Musto, pp. 2-3.
[119] See Jonnes (pp. 17-18) noting the disproportionate opiate use of middle class white women, whose doctors often prescribed the drug for "female problems".
[120] By 1912, "the framing of an antinarcotic law paralleled the widening possibilities open to Congress in the area of policing morals." Musto p. 10.
[121] Dulchinos, p. 13.
[122] Dulchinos, p. 13.
[123] Booth, p. 96, Dulchinos pp. 16-20.
[124] Abby nearly died giving birth to Fitz Hugh's older brother, who survived for less than a day (Dulchinos p.16).
[125] Dulchinos, p. 21.
[126] Booth, p. 96; Dulchinos, p. 21.
[127] Booth, p. 96; Dulchinos, p. 25.
[128] Dulchinos, p. 23.
[129] Dulchinos, p. 32.
[130] Dulchinos (p. 34) dates the time of Ludlow's first acquaintance with Mr. Anderson as "[s]ometime in 1853."
[131] F.H. Ludlow, *The Hasheesh Eater* (1857), pp. 15-17.
[132] Booth, p. 93: "The availability of psycho-active substances in America was widespread and there was no control over either their retail or their usage."
[133] See Dulchinos, p. 34.
[134] Dulchinos, p. 35. Contrary to popular belief, medical marijuana has been part of European culture for well over a millennium. "Partly because of its psycho-active potential, hemp was widely used throughout Europe as a folk medicine, just as it was in many other parts of the world. Much of this was as a result of the reliance both physicians and folk doctors placed upon Dioscorides' *Materia medica*, considered a central medical text until well into the seventeenth century. The

ailments treated with cannabis were many: it was used as an analgesic or anaesthetic to combat earache, toothache… rheumatism, arthritis, menstrual and labour pains, headaches and migraines and a large number of similar discomforts. It was also prescribed for epilepsy, inflammation, coughs, convulsions, fever and jaundice. A paste of cannabis flowers, olive oil and wax or animal fat was used as an ointment for open wounds" (Booth, p. 70). The medical uses of cannabis in 16th century Europe did not differ that greatly from its medical uses in the 21st century USA.

[135] Dulchinos, pp. 35-36. European associations between Arab culture ("the Prophet and his paradise") and cannabis use date back at least as far as the Crusades. By the time of the first Crusade, the use of hashish was already prevalent in Muslim society - Booth, p. 66.

[136] See note 102 on the introduction of cannabis to Africa by Arab traders. See also Booth (p. 55): "Probably under Arab influence, cannabis came to be regarded in Africa as a medicinal plant. It was used to fight malaria and blackwater fever, as an anesthetic during childbirth and as a cure for asthma, respiratory disease, dysentery and even anthrax."

[137] On February 17th, 1803 Coleridge wrote to his friend Thomas Wedgwood (who, like the poet, suffered from chronically poor health): "Last night I received a four ounce parcel letter by post… On opening it, it contained… a parcel, a small one, of Bang [sic] from Purkis… We will have a fair trial of Bang - Do bring down some of the Hyosycamine Pills - & I will give a fair Trial of opium, Henbane & Nepenthe. Bye the bye, I always considered Homer's account of the Nepenthe as a Banging lie." "Nepenthe" was the soporific drug given by Circe to Odysseus to make him forget about his homeland; popular myth has long equated nepenthe with hashish, though as the experienced Coleridge knows, the drug's description in the *Odyssey* more closely resembles laudanum than cannabis. As a side note, another of Coleridge's close friends, Humphry Davy, discovered nitrous oxide and supplied it to Coleridge's friends for recreational and creative purposes (Booth pp. 80-81).

[138] See Grotenhermen and Russo, pp. 42-44. "Born into a staunchly Roman Catholic family in Limerick, [O'Shaughnessy] attended the University of Edinburgh from the medical school of which he graduated in 1830. Three years later, at the age of twenty-four, he went out to work for the East India Company, being appointed assistant surgeon in Bengal and professor of chemistry at the Medical College in Calcutta" (Booth p. 109). See also Abel, pp. 105-06.

[139] O'Shaughnessy, W.B. "On the preparations of the Indian hemp, or Gunjah, *Transactions of the Medical and Physical Society of Bengal* (1838-1840).

[140] Later, a young Mark Twain would write similar articles for the San Francisco Examiner, the flagship newspaper of the Hearst media empire (Dulchinos, p. 181).

[141] See Grinspoon and Bakalar, p. 4; Werner, p. 65.

[142] The U.S. was once at the forefront of research into cannabis as a treatment for mental illness. As early as 1846, Dr. Amariah Brigham, the editor of the *American Journal of Insanity*, ran trials on the efficacy of cannabis on inmates of an insane asylum in Utica, New York (Booth, p. 114).

[143] Booth, p. 114. Sloman (p. 22) quotes from a listing in the 1851 *United States Dispensatory*: "Extract of hemp is a powerful narcotic (used here to indicate sleep-producing substance) causing exhilaration, intoxication, delirious hallucinations,

and, in its subsequent action, drowsiness and stupor, with little effect upon the circulation. It is asserted also to act as a decided aphrodisiac, to increase the appetite, and occasionally to induce the cataleptic state. In morbid states of the system, in has been found to cause sleep, to allay spasm, to compose nervous disquietude, and to relieve pain. In these respects it resembles opium; but it differs from that narcotic in not diminishing the appetite, checking the secretions, or constipating the bowels. It is much less certain in its effects, but may be preferably employed, when opium is contraindicated by its nauseating or constipating effects, or its disposition to produce headache, and to check the bronchial section. The complaints in which it has been specially recommended are neuralgia, gout, rheumatism, tetanus, hydrophobia, epidemic cholera, convulsions, chorea, hysteria, mental depression, delirium tremens, insanity and uterine hemorrhage."

[144] Dulchinos, p. 45.

[145] Werner, p. 65. According to Booth (pp. 112-13), Peter Squire of London was introduced to hashish by W.B. O'Shaughnessy himself.

[146] Grinspoon, p. 87.

[147] Dulchinos, p. 36.

[148] Dulchinos, p. 36.

[149] Ludlow, p. 20.

[150] *Ibid*, p. 24.

[151] Dulchinos, p. 37.

[152] Ludlow, p. 41. This description may not closely resemble the reader's own cannabis experiences or match very well with contemporary cannabis lore. While one may be inclined to attribute the descriptions to an overactive imagination or an intention to plagiarize DeQuincey, consider the argument of Campos (pp. 7-38, 155-180), in which he persuasively argues based on the theories of Norman Zinberg (*Drug, Set, and Setting*, Yale University Press, 1984) that subjective cannabis experiences may be to some extent culturally defined, so that the reader should not assume that a marijuana high in 21st America should be congruent with marijuana highs in other times and places. Grinspoon (pp. 56, 85-87) takes a more nuanced view, suggesting that many of the literary treatments of drug-induced states in the 19th century were influenced by the imaginations of their various authors absorbing ideas from reading DeQuincey; the previous work may have affected their subjective drug experience, their literary treatment of it afterwards, or both. Note also that Campos regrets (pp. 16, 20-21) the many confounding factors, not least of which was the tendency throughout history to mix hashish with opium and other drugs (see also Abel, p. 114 on this point).

[153] Ludlow, p. 41.

[154] *Ibid*, p. 43.

[155] Booth (p. 100) opines that "Ludlow's general academic studies seem not to have been adversely affected by his hashish 'habit'."

[156] Dulchinos, p. 42; Booth, p. 98.

[157] Dulchinos, p. 43. To this day, Union College students still traditionally sing Ludlow's "Ode to Old Union," which he wrote while a student there:

Let the Grecian dream of his sacred stream

And sing of the brave adorning
That Phoebus weaves from his laurel leaves
At the golden gates of morning.

But the brook that bounds thro' old Union's grounds
Gleams bright as a Delphic water,
And a prize as fair as a god may wear
Is a dip from our Alma Mater.

Then here's to thee, thou brave and free,
Old Union smiling o'er us,
And for many a day, as thy walls grow gray,
May they ring with thy children's chorus!

Ludlow set the words to the tune of a traditional drinking ballad (Dulchinos p. 69, Booth p. 101).

158 Dulchinos, p. 43. According to Booth (p. 98), "[f]or some time [in 1855], Ludlow had been going off the rails. His behaviour had been erratic, sometimes truculent and often supercilious or arrogant." Union College "had a reputation for handling gifted but errant students" (Booth p. 99).

159 *KA* was "the first undergraduate fraternity in the USA" (Booth, p. 101). See also Dulchinos, p. 54.

160 Dulchinos, p. 57; Booth, pp. 100-101.

161 Dulchinos, pp. 57-59.

162 Ludlow, p. 188. But not long before his trips started turning bad, Ludlow had read Edgar Allen Poe's masterful "The Pit and the Pendulum," which may have influenced Ludlow and through him an entire drug subculture for more than a century. See Dulchinos, p. 46; Booth, p. 100.

163 Dulchinos, p. 67: "Fitz Hugh had hallucinated the ultimate blasphemy against his father and his society, the death of God. (Nietzsche was at this time 12 years old.)" Booth, p. 102: "More and more, [Ludlow] interpreted his hashish dreams in religious or spiritual contexts. His suppressed guilt was getting the better of him."

164 There can be little doubt but that by the spring of 1856, Ludlow had become a problem user - he was taking up to four grams of hashish a day (Booth p. 102).

165 Dulchinos, p. 71.

166 Abel (pp. 111-12) declares that Ludlow "deliberately patterned" his work after DeQuincey's. DeQuincey himself announced in 1845 that he would soon write a sequel to *Confessions of An English Opium-Eater* based on a trial of "bang" he was about to receive. For some unknown reason, the self-administration of large amounts of cannabis failed to spark a prolific period in the writer's life; the project never materialized (Abel, pp. 103-04).

167 While circa 1855 there may have been a dearth of worthy literature written in the style of DeQuincey in English, it should be noted that the "Club des Hachichins" in Paris had already published some outstanding work in French - specifically about cannabis. Dumas' outstanding *The Count of Monte Cristo*, published 1844-45, contains an astonishing scene of a hashish trip after the eponymous protagonist encounters

an exotic Arabic character named Sinbad who introduces him to the drug. Gautier published "Le Club des Hachichins" in *La Revue des Deux Mondes* in February of 1846, in 1851 Gérard de Nerval published Voyage to the Orient, a well-received travelogue of the writer's travels through the Middle East which included descriptions of hashish use, and in the same year Baudelaire published an essay comparing wine and hashish. Still, Baudelaire's greatest accomplishment in drug literature, *Les Paradis Artificiels* ("The Artificial Paradise"), would not see the light of day until the year after Ludlow published *The Hasheesh Eater* in 1857.

[168] Dulchinos, p. 72. Booth, pp. 93-94: "Bayard Taylor appears to have been the first American to write about eating hashish. A very prolific author and journalist on the staff of the *New York Tribune*, Taylor... embarked on a journey to the Middle East during which, having a journalist's curiosity, he tried hashish in Egypt." Ironically, when Taylor published *Eldorado*, a travelogue of Mexico in 1854, he failed to mention cannabis, which was already quite prevalent there (Campos p. 70).

[169] Dulchinos, pp. 75-76.

[170] Dulchinos, p. 76.

[171] When Ludlow found it difficult to give up his habit, "[t]his made him realize he was addicted to hashish although he was not to know it was a psychological rather than physical addiction, such as he might have acquired from using opiates" (Booth p. 103). Yet it is known that Ludlow did in fact take morphine, although the exact date of his first regular use of the drug is not known (Dulchinos p. 198). He would have known the difference. Better to use the modern parlance more favored by physicians and diagnose Fitz Hugh Ludlow as dependent on cannabis - a more suitable term recognizing the complex and ill-understood relationship between the physiological and psychological aspects of the drug experience.

[172] Writing was not the only crutch Ludlow attempted to still his hashish habit. He "attempted to sublimate it with tobacco and, bizarrely, by blowing soap bubbles, the colours swirling on them reminding him of the marbling hues of his hashish hallucinations." Booth p. 104, Dulchinos p. 75.

[173] Dulchinos, p. 79.

[174] Ludlow borrowed the title from his friend Bayard Taylor; Taylor's piece in Putnam's which had so inspired Ludlow was titled "The Hasheesh Eater" (Booth, p. 95). The reference to the ancient Greek mathematician stems from "Ludlow's premise that Pythagoras... had used hashish with the members of his mystical brotherhood at Crotona" (Booth p. 104).

[175] Dulchinos, p. 87. Ironically, the Civil War also put an effective if temporary end to the American hemp industry, as the Union blockade of the Confederacy choked off the South's hemp trade and the abolition of slavery effectively ended its primary means of production (Grinspoon, p. 12).

[176] Dulchinos, p. 90.

[177] "In 1857 [shortly after the publication of *The Hasheesh Eater*], a Dr. John Bell, writing in the *Boston Medical and Surgical Journal*, commented on the proliferation of articles in the press appertaining to hashish and discussing its effects. He seemed to imply that there were, following in Ludlow's wake, an increasing number of his fellow American citizens experimenting with the drug. His implication was correct.

Cannabis was being used by some outside the strictly medical sphere" (Booth, p. 120). By 1883, "hasheesh houses" had appeared in the Hell's Kitchen neighborhood in Manhattan, where a patron could eat hash and in exotically-decorated lounges. The dens were frequented especially by the upper classes of the city (Booth pp. 122-23).
[178] Dulchinos, p. 87. Sloman (p. 26) quotes from an article by Mordecai Cooke in 1860: "Young America is beginning to use the 'bang,' so popular among the Hindoos… Lager beer and schnapps will give way for 'bang'."
[179] See Dulchinos, pp. 82-88. For some reason, the comparison of cannabis to opium is a recurrent theme in European/American history. Jan Anthoniszoon van Riebeeck, the first governor of the Dutch colony at Cape Town, wrote in 1652 that the dagga smoked by the local Hottentots seemed to have an effect similar to that of opium (Booth, p. 55). But Ludlow probably brought the comparisons on himself. "What Ludlow had set out to do was to write a polemic about hashish, just as Thomas DeQuincey had about opium" (Booth p. 105).
[180] Even contemporary authors have fallen into this trap. See, e.g., Booth (p. 108): "[H]ashish - a comparatively mild narcotic - came to be regarded as akin to the much more powerful and dangerously addictive opium, a misconception that has persisted to the present day."
[181] Compounding the problem, "any cannabis preparation left standing for a while tended to separate out, with the cannabis forming a residue at the bottom of the bottle. If it was not shaken vigorously, the patient receiving the last inch or so of the contents of the container tended to be heavily overdosed" (Booth p. 117).
[182] "[T]he medical use of cannabis was already in decline by 1890. The potency of cannabis preparations was too variable, and individual responses to orally ingested cannabis seemed erratic and unpredictable" (Grinspoon and Bakalar, p. 7). "As a medicinal agent, marijuana generally fell into disfavor before the turn of the century. For one, it was insoluble, and therefore couldn't be injected. So there were delays of up to three hours when administered orally. Second, there was tremendous difficulty in standardizing the dosage, as different batches showed great variations in potency. Also, there were variations among individuals in their response to the drug" (Sloman, p. 26).
[183] Fitz Hugh Ludlow spoofed "Mrs. Winslow's Soothing Syrup" in an adaptation of "Cinderella" for the stage (Dulchinos, p. 203).
[184] Musto p. 5: "perhaps 250,000" drug addicts in the U.S. in 1900. See also Bonnie and Whitebread, p. 8.
[185] Dulchinos, p. 248.
[186] Abel, p. 120.
[187] Dulchinos p. 248, quoting "What Shall They Do To Be Saved?" *Harper's Monthly*, August 1867.
[188] Dulchinos, p. 249.
[189] *Ibid.*
[190] *Ibid*, p. 250.
[191] *Ibid*, pp. 260-261. Ludlow also described dietary suggestions and even instructed on how to use electric shocks in emergency circumstances.

---

[192] Then there is the matter of substitution, which holds the promise of avoiding opium addiction before it starts: "Doctors were keen to prescribe [cannabis] because the only other effective painkiller they could offer was highly addictive opium and, it was soon realized, cannabis-based extracts were not physically addictive" (Booth p. 113). In 1904, G.F.W. Ewens, Superintendent of the Punjab Lunatic Asylum in Lahore, wrote of smoking *gunjah* that "it is universally believed that this habit has a great advantage over that of opium or alcohol or even tobacco taking in that it may be at any time relinquished without difficulty, and while I do not know whether this is absolutely true, I can certainly testify that no ill effects follow its sudden forcible stoppage against the will of the patient" (G.F.W Ewens, "Insanity Following the Use of Indian Hemp," *Indian Med. Gaz.*, 39 [1904], 402).
[193] *New York Times*, October 11th, 1868.
[194] *The Alantic Monthly*, November, 1868.
[195] It should be noted that some doctors did see in cannabis a lasting solution to the opium problem, although not in the way Ludlow envisioned. Rather, a growing coterie of doctors recognized the potential of hashish to replace opium in cases in which addiction or other complications from the more potent drug made it too risky to administer. As Dr. R.R. M'Meens reported to the Ohio State Medial Society in 1860, "its effects are less intense [than opium], and the secretions are not so much suppressed by it. Digestion is not disturbed; the appetite rather increased;... The whole effect of hemp being less violent, and producing a more natural sleep, without interfering with the actions of the internal organs, it is certainly often preferable to opium, although it is not equal to that drug in strength and reliability" ("Report of the Committee on Cannabis Indica," *Transactions*, Fifteenth Annual Meeting of the Ohio State Medical Society, Columbus, 1860). But H.A. Hare disagreed with M'Meens' characterization of cannabis as a weaker painkiller: "During the time that this remarkable drug is relieving pain, a very curious psychical condition sometimes manifests itself; namely that the diminution of the pain seems to be due to its fading away in the distance, so that the pain becomes less and less as a beaten drum was carried farther and farther out of the range of hearing" ("Clinical and Physiological Notes on the Action of Cannabis indica," *Therapeutic Gazette* 11[1887]: 225-26).
[196] Musto, pp. 79-80.
[197] Musto, pp. 79-82 includes a description of the Towns cure to make the stomach churn and the skin crawl.
[198] "A little less than ten percent returned to us for a second treatment, a reasonable presumption being that the ninety percent from whom we have never heard further that left our care had no need to consult with us a second time." Charles B. Towns, "The Physician's Guide for the Treatment of the Drug Habit and Alcoholism" (1914).
[199] Importation and Use of Opium, Hearings before the Committee on Ways and Means, House, 61st Congress, 3rd Sess., 11 Jan. 1911, pp. 50, 75-78. See also Musto pp. 47, 217.
[200] Jonnes (p. 19) notes that the *Gazette* was a "Parke-Davis-sponsored journal."
[201] Musto, p. 7.
[202] In keeping with medical tradition of the time (Jonnes, p. 19).

---

[203] After learning of some of the benefits of cocaine, Freud wrote to his sweetheart Martha, "We do not need more than one such lucky hit to be able to think of setting up house" (Jonnes, p. 19). Musto (p. 7) notes that Freud first learned of cocaine through reading American journals and that Uber Coca was abstracted in the *St. Louis Med. Surg. J.* 47:502-05 (1884), the same year as the book's publication in Vienna. By 1887 Freud's "wonder therapy" had already come under attack. S.A. Edminster et al., trans., *The Cocaine Papers* (Vienna: Dunquin Press, 1963). See also Siegfried Bernfeld, "Freud's Studies on Cocaine, 1884-1887, *Yearbook of Psychoanalysis* 10: 9-38 (1954-55) and Hortense Koller Becker, "Carl Koller and Cocaine," *Psychoan. Quart.* 32:309-73 (1963). Freud nicknamed his colleague Carl Koller "Coca Koller."
[204] Jonnes (p. 20) says Fleischl-Marxow embraced cocaine "like a drowning man."
[205] Jonnes, p. 22. But note that around the turn of the 20th century, even caffeine was occasionally controversial as a stimulant. An 1878 law in Mexico restricted its distribution through the herbolarias system (Campos, p. 186) and H.W. Wiley made headlines when he used his federal authority to seize a shipment of Coca-Cola under the theory that it contained dangerous levels of caffeine (Musto, p. 12).
[206] E.A. Birch, "The Use of Indian Hemp in the Treatment of Chronic Chloral and Chronic Opium Poisoning," *Lancet* 1 (1889): 625. In truth, the flurry of interest in cannabis as a treatment for chemical dependencies represented only a slice of the total medical establishment's treatment of the drug; Mikuriya reports that over a hundred papers on the benefits of cannabis were published in the journals of western medicine between 1840 and 1900. See Grinspoon and Bakalar, pp. 4-7; Booth, p. 114.
[207] J.B. Mattison, "*Cannabis indica* as an Anodyne and Hypnotic," *St. Louis Medical Surgical Journal* 61 (1891): 266.
[208] For a succinct summary of the technical reasons for cannabis' nadir as a medicine in the early twentieth century, see Grotenhermen and Russo, pp. 48-49.
[209] "[T]he work [of isolating the active principle of cannabis was] encouraged by the recent successes in isolating morphine, nicotine and caffeine. All these substances, however, were alkaloids but, as cannabis contains no active alkaloidal compounds… little progress was made" (Booth, p. 113).
[210] See table in Grotenhermen and Russo, p. 28.
[211] It is widely rumored but not conclusively proven that Reynolds gave Queen Victoria hashish to ease her menstrual cramps (Booth p. 114).
[212] Even though Reynolds reported that "[w]ith these precautions, I have never met with any toxic effects, and have rarely failed to find, after a comparatively short time, either the value or the uselessness of the drug," the personal attention required of the physician discouraged the titration technique's adoption (Booth p. 118).
[213] Booth, p. 115.
[214] Dulchinos, p. 268.
[215] *Ibid*, p. 270.
[216] Ludlow's death was probably due to alcohol more than any other substance. He began drinking especially heavily after his wife left him for his (former) best friend (Dulchinos p. 225).

---

217 Leslie Keeley, *The Morphine Eater*, Dwight, IL: C.L. Palmer & Co., 1881, p. 90.
218 According to Booth (p. 119), "[w]hat really put paid to cannabis as a viable medical product was the emergence of synthetic drugs created from the massive advances in chemistry made especially during the last thirty years of the nineteenth century. That these synthetics were often far more harmful than natural cannabis was considered by the way. It was the age of science, optimism and faith in scientific potential and infallibility overruling sensible judgement."
219 See Booth, p. 116; Musto, p. 5.
220 Jonnes, p. 13.
221 Booth, pp. 163-64; Bourrie, p. 44.
222 The stories from Anslinger's childhood come from Anslinger, Henry and Will Oursler. *The Murderers--The Shocking Story of the Narcotic Gangs* (1961). See also Jonnes, p. 91; Sloman, p. 31.
223 See Abel, p. 146.
224 King was, in many ways, merely mustering popular opinion. One newspaper wrote at around that time: "Opium and the Chinese, to the mind of your average newspaper reader, are inseparable" (H.W Morgan, *Yesterday's Addicts* [Norman: University of Oklahoma Press, 1974], p. 32).
225 According to Musto (p. 12), Wiley "developed the Agriculture Department's Bureau of Chemistry into an avid detector of unsavory manufacturing practices."
226 Other popular medicines containing morphine included "Battley's Sedative Solution" and "Godfrey's Cordial". Even major pharmaceutical companies like E.R. Squibb & Sons; Parke, Davis; Eli Lilly; and Bayer marketed over-the-counter and mail-order drugs containing opiates (Booth p. 116).
227 Such attitudes were by no means limited to alcohol. As Booth (pp. 154-55) notes, "drugs were identified in the public consciousness with foreigners and ethnic minorities who, in the strongly xenophobic and racist white portion of American society from with the administration and ruling class were drawn, were regarded with suspicion and already subjected to considerable social exclusion and repression. Drugs were, in short, deemed un-American."
228 For the cascade of statewide alcohol prohibitions in the early 20th century, see Bonnie and Whitebread, pp. 21-27.
229 Jonnes, p. 91.
230 See Booth, p. 155.
231 Jonnes, pp. 28-29. "Opium, perceived to be their [Chinese immigrants] insidious weapon of social destruction, was the subject of much vilification but the truth was different. Most Americans who could afford to buy a bottle of painkilling medicine had taken opium in one form or another for it was the most common analgesic available. The real cause of the hatred was racial" (Booth, p. 155).
232 "Despite a mushrooming opiate dependence problem arising from unrestrained distribution within the medical system, it was the 'street' use of opiates and cocaine which accelerated professional and public interest in their habit-forming properties. Indeed, legislative attention (and medical understanding of drug properties as well) seems to have occurred only after each new drug achieved a significant degree of 'street' use, especially when that use was identified with the poor, racial minorities and criminal classes… These early opium prohibitions, the first drug legislation to

criminalize the consumer for his indulgence, clearly had more to do with the drug's users than with the drug itself" (Bonnie and Whitebread, pp. 13-14). The judiciary at the time recognized the clearly racist intentions of these laws: "Smoking opium is not our vice, and therefore, it may be that this legislation proceeds more from a desire to vex and annoy the 'Heathen Chinee' in this respect, than to protect the people from the evil habit" (*Ex Parte Yung Jon*, 28 F. 308 [D. Ore. 1886] at 312.
[233] Jonnes, p. 25; Bonnie and Whitebread, p. 15. Musto (p. 22) estimates that sales of patent medicines dropped by a third after 1906. On August 22nd, 1908, Wiley boasted to Bishop Brent that "[a]lmost unanimous opinion that there has been a decrease from 25% to 50% in the sales of patent medicines containing opiates since the Pure Food Law went into effect."
[234] Jonnes, p. 40.
[235] See Musto, p. 24.
[236] Jonnes, p. 43: "Helping China was an opportune means to promote American interests." Musto, p. 4: "Indeed, a prime reason for calling the International Opium Commission was to mollify China's resentment of treatment of Chinese in the United States."
[237] Musto (p. 30) sums up the moralist basis of Bishop Brent's stance: "Did narcotics have a value other than as a medicine? No: unlike alcohol they had no beverage or caloric value. Should such substances be permitted for casual use? No: there was no justification, since there was the possibility only of danger in narcotics for nonmedicinal uses. Therefore recreational use of narcotics should be prohibited, their traffic curtailed on a world scale, and a scourge eliminated from the earth."
[238] Musto p. 31. Jonnes (p. 42) describes Wright as "a dashing man with bristling mustachios and a fervid manner."
[239] After the end of the first conference at The Hague, Bishop Brent had grown so exasperated with Wright's diplomatic manner that he wrote to him urging his retirement from the international stage. Wright refused (Jonnes, p. 46).
[240] Bonnie and Whitebread, p. 16. Jonnes, p. 44: The lack of federal anti-drug laws was "completely consonant with the nature of American governance then, for states at this period still retained exclusive police powers. That being the case, it was natural that American restrictions on drug use were expressed through a patchwork of municipal and state laws. Nonetheless, this presented a diplomatic problem."
[241] Booth, pp. 161-62.
[242] Musto (p. 10, ft. 24): "Wiley and the drug trades… cooperated to perfect an antinarcotic law based on the interstate commerce powers of the Constitution after it became apparent that pure food advocates feared any tampering with the Pure Food and Drug Act through amendment." *PHSR*, 2 Nov. 1908
[243] Musto pp. 34-35. The young American Medical Association campaigned vigorously for the ban (Jonnes, p. 29).
[244] Musto, p. 39.
[245] *Ibid*, p. 41.
[246] Jonnes, p. 45.
[247] Musto, pp. 41-42.

---

[248] The conflict between federalist principles and national drug policy elicited recognition early in U.S. history; as early as 1913 the AMA defensively editorialized that drug dangers "convince the most ardent advocate of states' rights that legislation regulating the sale of all dangerous habit-forming narcotics should be national in scope and absolutely uniform throughout the country." *JAMA* 60:1364 (1913)
[249] Hamilton Wright, "Report International Opium," in *Opium Problem: Message*, p. 45.
[250] David Musto, one of the country's foremost authorities on the history of American drug policy, puts the matter succinctly (p. 7): "evidence does not suggest that cocaine caused a crime wave but rather that anticipation of black rebellion inspired white alarm." A bout of historical karma may be at play; in a story eerily similar to that of cannabis's introduction to the New World, the Spanish conquistadors noted the use of coca leaves by indigenous Peruvians and made its use mandatory among the indigenous slaves of the silver mines in the 15th century to increase productivity and decrease appetite (Musto p. 8). As with cannabis, the colonizing culture introducing a drug developed an irrational and almost entirely unfounded phobia of its use by the colonized culture underfoot.
[251] Wright, "Report International Opium," pp. 48-49.
[252] *Ibid.* See also Booth, pp. 156, 162. Wright made the same argument in a pamphlet he published in 1910. See also Jonnes, p. 44.
[253] Musto (p. 151) credits Terry's clinic as what "may have been the first one established in the United States by a government agency, but it is likely that some arrangements for collective care of a locality's addicts had been made before that time."
[254] See Musto, pp. 97-100. The idea was not new in 1912; in 1902 E.G. Eberle had proposed a means-tested variation of the model Terry would use over a decade later. E.G. Eberle, "Chairman's Address, Minutes of the Section on Education and Legislation," *Proc. APhA* 50:550-61 (1902).
[255] But see Musto (p. 152) noting "substantial variations in clients' age and sex ratios." For very detailed data on the rates of addiction in the US over time, see Bonnie and Whitebread pp. 8-9, and especially footnote 11.
[256] New York state's "Whitney Acts" of 1917 and 1918. After brief and failed experiments with punitive anti-drug legislation championed by Charles Towns, New York enacted more compassionate legislation which allowed doctors and patients to decide the best addiction treatments. While the measures succeeded in relieving congestion in state criminal courts and reducing the influence of illegal distribution networks, it also allegedly contributed to the fortunes of a handful of unscrupulous doctors (Musto pp. 109-20). New York City also experimented with maintenance therapy, opening a clinic on Worth Street where morphine addicts could get treatment. Unlike the Jacksonville clinic, however, Worth Street aimed principally to cure and not study the addict population. See Musto pp. 140-41; Jonnes, pp. 53-54.
[257] See Musto, pp. 100-02.
[258] For summaries of three contemporaneous studies on the same theme, see Jonnes, pp. 49-50. Then there is the case of the state of Georgia, which after

reviewing admissions of 2,100 blacks to a Georgia asylum from 1909-1914, discovered to its surprise that only two had been cocaine users. E.M. Green, "Psychoses Among Negroes: A Comparative Study," *J. Nerv. Ment. Dis.* 41:697-708 (1914). Green believed that a simple lack of funds prevented "Negroes" from using cocaine at the same incidence as whites. But E.H. Williams replied in *Medical Record* (85:247-49 [1914]) that many examples of cocaine-addicted blacks could be found in the state's jails. Musto comments that "Williams in these writings does not seem so much anti-Negro as anti-Prohibition and uses the stories of cocainized Negroes to show what might happen if alcohol were not available" - p. 8, ft. 20.
[259] Another Southern maintenance clinic - this one in Shreveport, Louisiana - garnered general accolades as a well-known success story in public health (Jonnes, p. 56).
[260] Musto, p. 48.
[261] Booth, p. 163; Bonnie and Whitebread, p. 41; Abel, p. 128; Musto p. 218.
[262] Booth (p. 142) reports that "the Italian delegates asked for cannabis to be included in the discussions but they did not wait for the result of their request, leaving as they did after the opening sessions." South Africa, vís a vis its British controllers, had come to fear that cannabis would make their mineworking slaves both lazy and violent, two states not deemed mutually exclusive at the time. In the end: "the Conference is of the unanimous opinion that it is advisable to study the question of Indian hemp from the statistical and scientific point of view, with a view to regulating its abuses" (quoted in Booth, p. 142) - apparently ignorant of the British Indian Hemp Drugs Commission conducted 18 years earlier; it is regarded as the most thorough study ever conducted on the question (Booth p. 140, Campos pp. 17-18).
[263] Musto, p. 61. Harvey Washington Wiley also saw his career end in disappointment after he enraged president Roosevelt through his criticism of saccharin in 1912. See Musto, pp. 12-13.
[264] Musto p. 54. Bonnie and Whitebread (p. 16) have a succinct description of the Harrison Act's provisions.
[265] See Musto p. 6: "the story of the Harrison Act's passage contains many examples of the South's fear of the Negro as a ground for permitting a deviation from strict interpretation of the Constitution."
[266] Kremer and Urdang's *History of Pharmacy*, 3rd ed., revised by Glenn Sonnedecker (Philadelphia: Lippincott, 1963); R. Shryock, *The Development of Modern Medicine* (Philadelphia: Univ. of Pa. Press, 1947); M. Fishbein, *A History of the American Medical Association*, 1847-1947 (Philadelphia: Saunders, 1947); R. Stevens, *American Medicine and the Public Interest* (New Haven: Yale University Press, 1971); J.M. Burrow, *AMA: Voice of American Medicine* (Baltimore: Johns Hopkins Press, 1963).
[267] Musto (p. 14): "The public's fear of addicts and minority-group drug users might supply the powerful motive force for legislation, but the law's final form would await the approval of the institutional interests affected."
[268] Musto (pp. 11-12) identifies two distinct ideologies on the matter of drug addiction: those aimed critically against the greed of the for-profit patent medicine manufacturers, such as Samuel Hopkins Adams' "Great American Fraud" series in *Collier's* 1905-1907; and those aimed critically against the supposed immorality of

the drug user, often portrayed as a racial minority. "Both classes of reformers looked to federal legislation as the most effective weapon, and both tended to measure progress in the reform campaign by the amount of legislation enacted."

[269] The relationship between temperance advocates and cannabis had not always been antagonistic. In the 19th century, "[s]ome women's temperance societies advocated for [hashish] in place of alcohol: their reasoning was that drunks hit their wives, but cannabis users did not" (Booth, p. 121).

[270] Robinson's "facile schoolboy humour" (Booth, p. 133) may have also prejudiced readers against the subject matter of his essay.

[271] See Bonnie and Whitebread, pp. 48-51.

[272] Musto, p. 59-60.

[273] Besides, "[n]ot even the reformers claimed, in the pre-World War I hearings and debates over a federal antinarcotic act, that cannabis was a problem of any major significance in the United States" (Musto pp. 216-17).

[274] See Musto, p. 61.

[275] "More About the Harrison Bill," *JAPhA* 3:1-4 (1914).

[276] Editorial *JAPhA* 4:4-8 (1914). Beal believed strongly in the ability of the pharmaceutical industry to regulate itself, calling for internal measures as early as 1901 and warning that "[i]f pharmacists do not take up and deal rigorously with [rising rates of addiction to cocaine and morphine], they will be dealt with by the general public, and in a way not likely to be altogether agreeable to the pharmacists." J.H. Beal, "Report on Pharmacy Legislation," *Proc. APhA* 49:460-64 (1901).

[277] Musto, p. 64.

[278] "In retrospect, it is amazing that the Harrison Act in no way spelled out - despite all the years of dickering among the druggists, doctors, and politicians - one highly critical issue: What exactly did the new law expect of those most affected, the drug addicts?" Jonnes, p. 50. See also Bonnie and Whitebread, pp. 17-18.

[279] Musto, pp. 121-22, 126; Jonnes, p. 51; Bonnie and Whitebread, pp. 18-20.

[280] Musto, p. 130.

[281] *Ibid*, pp. 131-34. Jonnes (p. 52) places part of the blame of the sudden drug hysteria on the shoulders of the Treasury, which in 1919 published a pamphlet making such absurd and alarmist claims as to estimate New York City's addict population at 300,000 (almost three times the likely *national* total).

[282] The Court was particularly sensitive to public opinion in the teens. As one contemporary source noted, judicial rulings on the reach of the federal government "could not be determined in advance by abstract reasoning. Hence, as each litigation arose, the judges could follow no rule but the rule of common sense, and the Police Power, translated into plain English, presently came to signify whatever, at the moment, the judges happened to think reasonable. Consequently, they began guessing at the drift of public opinion, as it percolated to them through the medium of their education and prejudices. Sometimes they guessed right and sometimes wrong, and when they guessed wrong, they were cast aside, as appeared dramatically enough in the Temperance agitation" (B. Adams, *The Theory of Social Revolutions* [New York: Macmillan, 1913], p. 94).

[283] The debate over alcohol prohibition probably helped to overshadow the debate over the Harrison Act, pushing matters of drug policy and constitutional limits temporarily out of the public mind. See Jonnes, pp. 47-48.
[284] As Musto notes, this "direct connection between the efficiency needed for the war effort and the inefficiency caused by narcotic use was made by the Treasury Department. Warning of narcotic use as a threat to the war effort appeared in the Commissioner of Internal Revenue's annual report for 1917." See also Marks, Janette, `"Narcotics and the War," *North American* 206:879-84 (Dec. 1917).
[285] Writing of Progressive principles, Bonnie and Whitebread (p. 10) remark that "[m]oral reform had been a strong undercurrent of nineteenth-century American life, and - as in the case of alcohol - every social ill attracted its own group of organized adversaries. As the century progressed and the institutional evils of industrialization and urbanization became apparent, this reform sentiment gradually turned to the government and the law to protect the moral fiber of the nation. The moral strength of the individual was no longer believed sufficient to counter corporate selfishness, political corruption, and urban degradation."
[286] Jonnes, p. 92; Sloman, p. 32. Musto (p. 210) implies that Anslinger's recovery of the Kaiser's field kit was no great feat but rather represented a failed mission.
[287] Musto, p. 211; Booth, p. 175; Jonnes, pp. 92-93.
[288] Booth, p. 175.
[289] Quoted in Sloman, p. 33.
[290] See Jonnes, pp. 73-79.
[291] Booth, p. 44: "The 1830s saw a rapid increase in the invention of hemp machinery, especially for the making of rope and hessian. In 1838, David Myerle of Philadelphia built a steam-driven factory at Louisville, Kentucky, based upon an invention by one Robert Graves of Boston, from whom he bought the patent. Then, in 1841, Andrew Caldwell of Lexington, Virginia invented a mechanized system in which raw hemp fibre [sic] was separated, spun into yarn and sewn into hessian sacking."
[292] Grinspoon, p. 13.
[293] For decades, the purveyors of hemp seed for bird and human consumption have taken advantage of a loophole in US law, allowing hemp seed which has been sterilized (Booth p. 187).
[294] "[I]n the mid-1800s… Kentucky farmers began to import Chinese rather than European seed, improving the quality of domestic hemp" (Bonnie and Whitebread, p. 3). China, which is still a leading producer of industrial hemp today, has been farming hemp for about 8,500 years (Small and Marcus, p. 284).
[295] See Small and Marcus, p. 288.
[296] Dewey and Merrill, 1916. See also Small and Marcus, pp. 294-95.
[297] The same report found that an acre sowed with hemp could produce more than four times the paper pulp than an equivalent acre of forest land. *Ibid.*
[298] Lyster H. Dewey, "Hemp," in USDA Yearbook of Agriculture 1913.
[299] See Ranalli and Venturi, pp. 2-3.
[300] See Booth, p. 185: "Many inventors, since the time of Thomas Jefferson, had come up with hemp processors but only the decorticator patented on 1 July 1919 by George W. Schlichten worked effectively at an industrial level." From Herer,

Chapter 2: "Henry Timken, a wealthy industrialist and inventor of the roller bearing got wind of Schlichten's invention and went to meet the inventor in February of 1917. Timken saw the decorticator as a revolutionary discovery that would improve conditions for mankind. Timken offered Schlichten the chance to grow 100 acres of hemp on his ranch in the fertile farmlands of Imperial Valley, California, just east of San Diego, so that Schlichten could test his invention. Shortly thereafter, Timken met with the newspaper giant E.W. Scripps, and his long-time associate Milton McRae, at Miramar, Scripps' home in San Diego."
[301] See Bourrie, pp. 23-26; Herer, Chapter 4.
[302] "Mexico at that time [circa 1900] was in a state of characteristic upheaval. The capitalist dictatorship of General Porfirio Díaz was unpopular and dissent was rife, the economy at best unstable" (Booth p. 158).
[303] According to Booth (p. 159), "[i]t was not [Villa's] first incursion north of the Rio Grande: Villa had been in the habit of raiding border towns and ranches to obtain supplies for some years."
[304] See Booth, pp. 158-59.
[305] "The Mexican marihuana folklore apparently made a deep impression on any American who came in contact with the drug or its alien users. Having no reason to suspect the veracity of such tales, law enforcement officials and local representatives… of the federal government agitated for state and federal legislation to combat the 'killer weed'" (Bonnie and Whitebread, p. 37).
[306] For the cascade of prohibitions sweeping Mexico between 1869 and 1896, see Campos, pp. 193-94.
[307] Bourrie, p. 44. A contemporary government report described El Paso as a "hot bed of marihuana fiends" ("Report of Investigation in the State of Texas," R.F. Smith [Washington, D.C.: Bureau of Chemistry, 1917]).
[308] See, for example, headlines from Hearst newspapers the *Los Angeles Examiner* "Murder Weed Found Up and Down Coast - Deadly Marihuana Dope Plant Ready For Harvest That Means Enslavement of California Children" (Nov. 5th, 1933) and the *San Francisco Examiner* "Dope Officials Helpless to Curb Marihuana Use" (Nov. 7th, 1933). "The Hearst chain were leaders in [printing lurid stories linking marijuana use and violence], but even the newspaper of record, the *New York Times*, reported these horror stories" (Sloman p. 30). The move reflected shifting attitudes about cannabis and darker-skinned races worldwide. In 1928 the British journalist Malcolm Muggeridge published an autobiography in which he described some of his college students in Cairo who took to smoking hashish: "They seemed to be faraway, lost in some dream of erotic bliss; a consequence, no doubt, in the case of many of them, of their addiction to hashish, widespread among the effendi class, and prevalent among the fellahin, especially among the ones who had moved into the towns" (Booth, p. 144).
[309] See Campos, p. 219.
[310] The notion that the smoking of cannabis causes insane and violent states is old and various in its origins. The most direct antecedent to Hearst's version came from Mexico itself, where the local folklore associated its use by indigenous and lower classes with racial degeneracy and violent insanity. Indeed, Hearst and another newspaper services like the Associated Press merely had to translate

---

sensational articles already appearing in Mexico City (see Campos p. 205: "[B]y the 1890s marijuana's reputation for causing madness and violence was well established in Mexico, and over the next two decades that reputation would be reinforced with sensational detail by a flourishing yellow press. It was from this atmosphere that the U.S. media began to pluck exotic, sensational stories of a new drug menace south of the border"). But note that the Mexican Herald, an English paper in Mexico City which Campos (p. 206) notes as especially influential to forming the American media narrative, confused "marihuana" with powerful hallucinogens like jimson weed from the start of its coverage. Such confusion may have contributed to cannabis's burgeoning identity as a 'loco weed' at this time (Campos 207-218).

Nor was Mexico the only source of such stories. A.T. Bryant, in his book *The Zulu People* (1967), wrote that Zulu warriors "under the exciting stimulation of [dagga] were capable of accomplishing hazardous feats." British folklore also held that dagga smoking before the battles of Isandhlwana and Rourke's Drift in 1879 gave the Zulu warriors sufficiently enhanced strength as to explain their unexpected ferocity on the battle field.

David Livingstone, famous throughout Europe for his journeys through Africa, wrote of the Bakota tribe's use of cannabis: "They like its narcotic effects, though the violent fit of coughing which follows a couple of puffs of smoke appears distressing, and causes a feeling of disgust in the spectator. This is not diminished on seeing the usual practice of taking a mouthful of water, and squirting it out together with the smoke, then uttering a string of half-incoherent sentences, usually in self-praise. This pernicious weed is extensively used in all the tribes of the interior. It causes a species of phrensy [sic], and Sebituane's soldiers, on coming in sight of their enemies, sat down and smoked it, in order that they might make an effective onslaught… Never having tried it, I can not describe the pleasurable effects it is said to produce, but the hashish in use among the Turks is simply an extract of the same plant, and that, like opium, produces different effects on different individuals. Some view every thing as if looking in through the wide end of a telescope, and others, in passing over a straw, lift up their feet as if about to cross the trunk of a tree." David Livingstone, *Missionary Travels and Researches in South Africa* (1858).

[311] Campos, pp. 161-63.

[312] Booth, p. 179: "It was Hearst, or more accurately his newspapers, which popularized the word marijuana… Prior to the 1930s, those who know of the drug, and were neither Mexican nor black, usually referred to it as hemp. Marijuana, as it gained common parlance, came to stand for a criminal substance, a foreign drug set to undermine mainstream American culture."

[313] Booth, p. 161: "Marijuana was labeled an alien drug, the fact that it was grown in America and had been an ingredient of patent medicines for decades was conveniently overlooked."

[314] Bonnie and Whitebread, pp. 37-42.

[315] See, e.g., C.M. Goethe of Sacramento, California: "Marijuana, perhaps now the most insidious of our narcotics, is a direct by-product of unrestricted Mexican immigration. Easily grown, it has been asserted that it has recently been planted between rows in a California penitentiary garden. Mexican peddlers have been

caught distributing sample marijuana cigarets to school children." NY Times, 15 Sept. 1935. See also Booth, p. 161: "The underlying reason for [El Paso's pioneering anti-cannabis ordinance] was not to prohibit cannabis but suppress the Mexicans. Once the potential of marijuana legislation for suppressing the migrants was realized, other cities and districts were quick to imitate it." Abel (p. 134) points out that these laws and newspaper articles came out during a time in which Mexicans were generally disfavored by the broader American public.
[316] There is even some evidence to suggest that government prohibitions of opium caused greater street demand for heroin in the early 20th century (Jonnes, pp. 110-11).
[317] This book's discussion of Rothstein's drug career owes a large debt to the treatment of the same subject matter by Jonnes, pp. 72-86.
[318] Jonnes, p. 75.
[319] Rothstein operated early in the vacuum created by the US zero-maintenance paradigm; within a decade Washington, D.C. would be shutting down government-regulated maintenance clinics in Mexico as well (Campos, p. 226).
[320] Jonnes, p. 78.
[321] The cultural definition taken on by the word "narcotic" is quite instructive of the times. "The most important feature of this initial prohibitory phase [between 1913 and 1931] is that marihuana was inevitably viewed as a 'narcotic' drug, thereby invoking the broad consensus underlying the nation's recently enunciated antinarcotics policy. This classification emerged primarily from the drug's alien character. Although use of some drugs - alcohol and tobacco - was indigenous to American life, the use of 'narcotics' for pleasure was not. Evidently, drugs associated with ethnic minorities and with otherwise 'immoral' populations were automatically viewed as 'narcotics.' The scientific community shared this social bias and therefore had little interest in scientific accuracy. From this instinctive classification of marihuana with opium, morphine, heroin, and cocaine flowed the entire set of factual supports on which narcotics prohibition rested" (Bonnie and Whitebread, pp. 51-52).
[322] Small and Marcus, p. 1.
[323] See Herer, Chapter 2.
[324] "New Billion Dollar Crop," *Popular Mechanics*, Feb. 1937. For an updated list of potential industrial applications, see Small and Marcus, p. 285.
[325] See Booth, p. 184; Small and Marcus, p. 299.
[326] Herer, Chapter 4.
[327] See Jonnes, pp. 81-82.
[328] Col. Nutt had been a pioneer of anti-maintenance policies, deciding as early as late 1919 to close down maintenance clinics while head of the Treasury's Narcotic Division (Musto pp. 148, 206-08; Booth p. 176). Before the investigation, he enjoyed a generally forthright reputation (Musto p. 183).
[329] Jonnes, p. 88. The Stephen Porter Narcotic Farm Act of 1929 made history by establishing the first federally-operated addiction treatment facilities in US history. The two "farms" at Lexington, Kentucky and Fort Worth, Texas never had a high treatment success rate, but as Musto (p. 206) notes, neither did any other treatment commonly used in the 1930s. Bonnie and Whitebread (p. 56) note that this law was

the first federal legislation to classify cannabis as a narcotic, lumping it together with opium and morphine.
330 Jonnes, p. 85.
331 Jonnes, p. 84.
332 Booth, p. 176.
333 See Jonnes, p. 42. Her politically-connected relatives included her father William Drew Washburn, who represented Minnesota in both the House and the Senate and uncles Cadwallader, Israel, and Elihu Washburn, who were elected to Congress from Wisconsin, Maine, and Illinois, respectively. *NY Times* obit. Feb. 14, 1952.
334 Elizabeth Wright "very much expected to have a job in the new narcotics bureau" (Jonnes, p. 89).
335 Jonnes, p. 89.
336 Musto, p. 209.
337 Jonnes, p. 85.
338 Musto, pp. 208-09; Booth, p. 185. Booth (p. 174) calls Anslinger "probably the most important player in the history of both American and international anti-narcotics law enforcement and legislation."
339 "[Mellon] had vested interests in seeing hemp suppressed as he was a major shareholder in Gulf Oil and a huge coal mining concern in Pennsylvania, not to mention a number of utility companies" (Booth, pp. 185-86).
340 See Musto, p. 209.
341 For a discussion on the bureaucratic rivalry between Anslinger and Hoover, see Booth, p. 177.
342 Jonnes, p. 90.
343 Musto (pp. 211-12): "Anslinger would have made the purchase of alcohol for nonmedical consumption a violation, and for the first conviction would set a penalty of a fine of not less than $1,000 and imprisonment for not less than six months. For a second or subsequent offense the fine would be between $5,000 and $50,000 and imprisonment for two to five years. In his view these penalties would put teeth into the law and greatly discourage violations… There is no evidence that during Anslinger's tenure as commissioner he ever changed his mind that the most effective way of gaining public compliance with a law regulating a dangerous drug was a policy of high fines and sever mandatory prison sentences for first convictions."
344 Booth, p. 180: "The Depression caused a considerable fall in tax revenue and government spending plummeted. The FBN budget was substantially cut."
345 Musto, p. 212; Booth, p. 177.
346 Internally, the Bureau often took a different position. See "Report by the Government of the United States of America for the Calendar Year Ended December 31, 1931: On the Traffic in Opium and Other Dangerous Drugs," Federal Bureau of Narcotics (GPO, 1932), p. 51: "This abuse of the drug is noted among Latin-American or Spanish-speaking population. The sale of cannabis cigarettes occurs to a considerable degree in States along the Mexican border and in cities of the Southwest and West, as well as in New York City, and, in fact, wherever there are settlements of Latin Americans. A great deal of public interest has been aroused by newspaper articles appearing from time to time on the evils of

the abuse of marijuana or Indian hemp, and more attention has been focused upon specific cases reported of the abuse of the drug than would otherwise have been the case. This publicity tends to magnify the extent of the evil and lends color to an inference that there is an alarming spread of the improper use of the drug, whereas the actual increase in such use may not have been inordinately large."

[347] Jonnes, p. 128. Anslinger continued to insist on state solutions to the "marihuana menace" as late as January of 1937. NY Times, 3 Jan. 1937.

[348] "It is clear that the... Prohibition Bureau [the immediate predecessor to the FBN] would have preferred to eliminate the smoking of marihuana... But there was a *legal* objection to the proposed legislation. Opinion was virtually unanimous that the inclusion of cannabis in the Harrison and Import and Export acts was illogical, possibly unconstitutional, and might even endanger the entire federal legislative scheme" (Bonnie and Whitebread, p. 60).

[349] "On the night of January 18, 1917, Brigadier General José María Rodriguez, the acting head of the Federal District's Superior Sanitary Council and personal physician of Venustiano Carranza, rose before the delegates to the Constituent Convention at Querétaro [to draft a new constitution for Mexico] and argued passionately that, for the sake of Mexico's national survival, a certain 'tyranny' of a newly constituted sanitary council should be written into the revolutionary charter" (Campos, p. 197). The convention found Rodriguez persuasive, and the following month approved a new constitution granting strong federal control of drug policy.

[350] Campos, p. 198.

[351] Musto p. 10.

[352] "Soon after the Senate confirmed his appointment on 18 December 1930 Commissioner Anslinger decided that the bureau's first major project should be an active involvement in the drafting of the Uniform State Narcotic Drug Act" (Bonnie and Whitebread, p. 67). Some have suggested that Anslinger was motivated by fiscal considerations. "In order to boost his organization, Anslinger had to find a new target - a new drug menace - upon which to peg a budget increase. Although he had previously given marijuana little thought and deferred putting it under federal legislation, he now set about... pushing for marijuana to be included as a dangerous drug alongside opiates and cocaine in the Uniform State Narcotic Acts" (Booth, p. 180).

[353] Early in his career, Dr. Woodward had headed the AMA's Bureau of Legislation. See Musto, p. 57.

[354] See Bonnie and Whitebread, pp. 80, 85.

[355] Bonnie and Whitebread, p. 82.

[356] Sloman, p. 39: "One pharmacist railed, 'Absolute rot. [Prohibition] is not necessary. I have never known of [cannabis's] misuse.'"

[357] Bonnie and Whitebread, p. 65.

[358] See Campos, pp. 18-19; Bourrie, pp. 41-42; Bonnie and Whitebread, pp. 132-35; Abel, pp. 130-131.

[359] J.F. Siler et al, "Marihuana Smoking in Panama," *Milit. Surg.*, 73 (1933). See also Grinspoon, p. 19; Booth, p. 160.

[360] *1930 of the National Conference of Commissioners on Uniform State Laws and Proceedings* (Baltimore: Lord Baltimore Press, 1930), p. 493.

---

[361] Bonnie and Whitebread, pp. 82-84.
[362] Quoted in Sloman, pp. 38-39; Bonnie and Whitebread, p. 64.
[363] "Anslinger's purpose in going to the press at this time is quite apparent. He had decided that there was no legitimate medical use of the drug, that allowing such use would merely open a gigantic loophole in the law, and that, accordingly, the essential first step was for each state to enact a total ban on cultivation, sale, and possession of marihuana. To implement this, the FBN had just proposed such a provision to the Conference of Commissioners for Uniform State Laws. The bureau proposal had been defeated, however, by the medico-pharmaceutical interests. Consequently, Anslinger took his case to the press" (Bonnie and Whitebread, p. 77).
[364] Bonnie and Whitebread, pp. 86-87.
[365] Officially, the FBN remained neutral in the Woodward affair (Bonnie and Whitebread, p. 88).
[366] "After considerable comment, it was decided to eliminate Section 12 (Cannabis) and leave it to the Conference of Commissioners as to whether it should be included under the general provisions of the Act" (Report of the Preliminary Conference Held in Federal Reserve Board Conference Room, Treasury Department, to discuss the Fifth Tentative Draft of the Uniform State Narcotic Law [Washington, D.C., 15 Sept. 1932], p. 23).
[367] Bonnie and Whitebread, pp. 88-91.
[368] Booth, p. 185. Herer (Chapter 2) alleges that Oliphant received significant pressure from the DuPont Company to favor its synthetic varnish over hempseed oil industry, which in 1935 accounted for 58,000 tons of naturally-derived varnish.
[369] "A large part of the bureau's activity consisted of intensive lobbying in each legislature before which the [Uniform State Narcotic] act was pending. Anslinger instructed his district supervisors and local agents to campaign actively with state legislators for the passage of the act, urging them to make as many speeches and public appearances as possible to marshal public support... In an important strategic decision, the commissioner suggested that his three hundred agents work directly with legislators" (Bonnie and Whitebread, p. 95). Anslinger even co-wrote a law review article with his legal counsel A.L. Tennyson urging the national Bar to adopt the Uniform Acts (Anslinger, "The Reason for Uniform State Narcotic Legislation," *Georgetown Law Journal* 21 [1932]).
[370] Bonnie and Whitebread, pp. 103-12. Anslinger's FBN made canny use of the WCTU and the General Federation of Women's Clubs, who acted as a surrogate public relations army to defuse the Commissioner's detractors when expedient. See Musto, p. 214.
[371] "Anslinger won the day, aided by the press, especially the Hearst newspapers which whipped up public opinion [against marijuana]" (Booth, p. 180).
[372] Sloman, p. 47; Bonnie and Whitebread, pp. 95-97.
[373] "Anslinger, forty-two, would complain privately that his nerves felt jangled, and in the fall of 1934 he was horrified to realize his hair was falling out" (Jonnes, p. 104). His doctor diagnosed him with "a form of nervous strain incident to his professional duties."
[374] See Bonnie and Whitebread, pp. 97-100.

---

[375] Anslinger and Cooper, "Marihuana: Assassin of Youth," *American Magazine,* July 1937.
[376] "One clear indication of Bureau influence in the preparation of journalistic articles can be found in the recurrence of certain atrocity stories first reported by the Bureau" (H.S. Becker, *Outsiders: Studies in the Sociology of Deviance* [New York, 1963], p. 141). Indeed, of 17 articles condemning marihuana in the popular press published between July 1937 and June 1939, 10 explicitly acknowledged the help of the Federal Bureau of Narcotics. See also Grinspoon, pp. 17-20, 323-26; Jonnes, p. 129; Sloman, pp. 35-37. A short excerpt from "Marihuana: Assassin of Youth" provides a good feel for the article: "The sprawled body of a young girl lay crushed on the sidewalk the other day after a plunge from the fifth story of a Chicago apartment house. Everyone called it suicide, but actually it was murder. The killer was a narcotic known to America as marijuana, and to history as hashish. It is a narcotic used in the form of cigarettes, comparatively new to the United States and as dangerous as a coiled rattlesnake." Ironically, Anslinger would report in 1970 that he had always felt that heroin was much more dangerous than marijuana (Sloman, p. 80).
[377] To publicize the new evil, Anslinger went, as usual, to his friends at the Hearst newspaper chain (Bonnie and Whitebread, p. 100).
[378] See Booth, p. 181. Acquiring adequate funding became a never-ending task for Anslinger. During the lean thirties, the FBN had to squeak by on annual budgets of $1.1 to $1.3 million (Musto p. 214). Jonnes (p. 103) suggests that in 1933, Anslinger believed his entire Bureau to be in jeopardy of dismantling.
[379] "Crazed Youth Kills Five of Family With Ax In Tampa," *Tampa Times,* October 18th, 1933.
[380] Sloman, p. 62; Abel, pp. 132-33.
[381] See Booth, pp. 183-84.
[382] See Bonnie and Whitebread, pp. 120-23. The famous "migratory bird case," *Missouri v. Holland* (1920), laid the constitutional foundation for an expansion of federal power not otherwise authorized by the Constitution if a foreign treaty required the federal government to use such a power to enforce a certain law. For a time, therefore, an international treaty requiring federal cannabis prohibition appeared an attractive option for heading off constitutional challenge.
[383] The first Geneva Conference, held in 1924, had passed sweeping agreements on the international control of cocaine and opiates but passed only tepid resolutions on cannabis, restricting only the export of hash extract for other than medical or scientific purposes (see Booth, p. 142; Abel, p. 148). Anslinger sincerely wanted stronger restrictions on cannabis the second time around.
[384] Booth, p. 186.
[385] Sloman, p. 49.
[386] Musto, p. 222; Werner, p. 66.
[387] Bonnie and Whitebread, p. 126. "Hester [C. Hester, Assistant General Counsel for the Treasury] explained [to Congress] that by differential taxation of registered and unregistered marihuana transactions, the federal government could restrict use without impinging on the rights of drug regulation reserved to the states according to the Tenth Amendment of the Constitution. The Tax Act provided that: (1) all

individuals using the plant for defined industrial or medical purposes must register and pay a tax of $1.00 per ounce; (2) individuals using marihuana for purposes undefined by the Act must pay a tax on unregistered transactions of $100 per ounce; and (3) any individual failing to comply with the above regulations was subject to penalties for tax evasion... or a fine of not more than $2,000 and/or a prison sentence of not more than five years" (Grinspoon, p. 21). Grinspoon and Bakalar, p. 8: "[T]he Marihuana Tax Act... was put in the form of a revenue measure to evade the effect of Supreme Court decisions that reserved to the states the right to regulate most commercial transactions. By forcing some marihuana transactions to be registered and others to be taxed heavily, the government could make it prohibitively expensive to obtain the drug legally." See also Grinspoon, p. 11.

As usual, the US was following behind its neighbor to the south. Mexico had already passed federal anti-marijuana legislation in 1920, the prohibitionary law titled "Dispositions on the Cultivation and Commerce of Products that Degenerate the Race" (Campos p. 181; Bonnie and Whitebread, p. 36). In light of the US law cloaking a prohibition in the form of a tax, compare to ironically inverted regulations passed by British Parliament in India after wresting control from the East India Company in 1784: "In 1790, one of the commodities they taxed was cannabis. This was followed three years later by a regulation licensing the production and sale of... Indian hemp drugs. The level of taxation was based upon the potency of the drug. As with so many taxes, it was claimed the imposition was to control consumption: at the same time, alcohol and tobacco were also taxed, supposedly for the same reason. In truth it was really aimed at raising revenue. Parliament had little genuine interest in the welfare of the native Indian population" (Booth p. 75).

[388] Marihuana questionnaire filled out by Dr. Walter Treadway, Papers of Harry J. Anslinger, box 6. Deposited in the Pennsylvania Historical Collections of Pennsylvania State University.

[389] The Treadway affair was no isolated incident. As Musto (p. 225) puts it, "[t]he Treasury Department collected and considered scientific and medical opinion prior to the Tax Act hearings, but the desire to present a solid front when the department appeared before the committees of Congress caused the officials to ignore anything that qualified or minimized the evils of marihuana."

[390] The Treasury's legal counsel Alfred Tennyson objected to broad definitions which would prohibit every form of the cannabis plant: "We have here some other uses... There is a use for fiber, for birdseed, and for oil in the varnish industry. These people will probably come in and complain about what they consider a foolish attempt to control if we try to make this all-inclusive." But Anslinger stood firm. "We might be in a bad position if we eliminated the stalks [from the list of prohibited constituents] and later found [cannabinoids] to be present in them." At the suggestion that only the active hashish resin of the plant be proscribed, Anslinger insisted that his own agents would be confused by field drug tests (Sloman, pp. 52-54).

Anslinger had his way. The final draft of the Tax Act defined "marihuana" as: "all parts of the plant Cannabis sativa L., whether growing or not; the seeds thereof; the resin extracted from any part of such plant; and every compound, manufacture, salt, derivative, mixture, or preparation of such plant, its seeds, or resin; but shall not include the mature stalks of such plant, fiber produced from such stalks, oil or cake made from the seeds of such plant, any other compound, manufacture, salt, derivative, mixture, or preparation of such mature stalks (except the resin extracted therefrom), fiber, oil, or cake, or the sterilized seed of such plant which is incapable of germination" (Marihuana Tax Act of 1937 § 1(b), ch. 553, 50 Stat. 551).

[391] Sloman, p. 58.

[392] Populist movements in Latin America had been partly responsible for a drag on returns in the American oil industry; as early as 1919, US oil concerns began to lobby Congress to prevent the implementation of parts of the 1917 Mexican constitution which allowed regulation of foreign business interests operating in Mexico (Booth, p. 223). The oil lobbyist told Congress that "large proportions of Mexicans, officers as well as men, are dope fiends. They smoke mariguana, which is made from the loco weed familiar to cattle men in the Southwest [actually astragalus] which has an effect like hasheesh. They will not go into battle without a dose of mariguana which imparts a sort of false courage" (Subcommittee on the Committee on Foreign Relations of the United States Senate, Sixty-Sixth Congress, 457-64).

[393] DuPont Chemical Company, 1937 Annual Report to Shareholders

[394] Many readers balk at the conspiracy theory-centered thesis of Jack Herer's hugely influential *The Emperor Wears No Clothes*, full of assertions like "a series of secret meetings were held." Other opinions diverge on the likelihood of a DuPont-centered conspiracy to crush a nascent industry. See, e.g., William S. Burroughs: "Harry J. Anslinger becomes comprehensible only as part of a conspiracy" (Sloman, p. *xi*); Booth (p. 184): "There is much of the conspiracy theory about [Herer's] premise but it is not at all outlandish."

[395] "The narcotics bureaucracy had no definitive scientific study of the effects of marihuana to present to the Congress. Even so, one might have thought the Treasury Department would have submitted a synthesis of available scientific information, or perhaps would have summoned a number of private investigators of the government's own public health experts to testify about the drug's effects. None of these things were done" (Bonnie and Whitebread, pp. 154-55). "A reading of the hearings that preceded the passage of the [Marihuana Tax Act] demonstrates quite clearly how little empirical data was found to support [the] legislative judgment [that marijuana is a 'harmful substance']" (Grinspoon, p. 20). According to Bonnie and Whitebread (p. 127), the first time the FBN convened a conference to determine the effects of marijuana as a drug was two weeks after they criminalized it.

[396] *Taxation on Marihuana, Hearings on H.R. 6385*, p. 30.

[397] See, e.g. the transcript from the Senate hearings on the Tax Act (Senate Finance Committee Subcommittee, *Hearings on H.R. 6906*, 75th Cong., 1st sess., 1937):

> Mr. ANSLINGER. [after describing a gruesome murder] We have many cases of this kind.
>
> Senator BROWN. It affects them that way?

Mr. ANSLINGER. Yes.
Senator DAVIS (viewing a photograph presented by Mr. Anslinger). Was there is this case a blood or skin disease caused by marihuana?
Mr. ANSLINGER. No, this is a photograph of the murdered man, Senator. It shows the fury of the murderer.
Senator BROWN. That is terrible. [Anslinger proceeds to describe other cases including the Licata case.]

398 *Ibid.*

399 *Taxation on Marihuana, Hearings on H.R. 6385*, pp. 48-52.

400 Booth, p. 187; Abel (p. 156) records Woodward's protests that the drafting of the Marihuana Tax Act was apparently kept secret from the AMA.

401 *Taxation on Marihuana, Hearings on H.R. 6385*, p. 50. For this and more quotes from the Congressional hearings, see Sloman, pp. 75-79.

402 Bonnie and Whitebread, p. 165; Abel, pp. 153-54.

403 Representative Dingell cut off Woodward's testimony thus: "You are not cooperative in this. If you want to advise us on legislation you ought to come here with some constructive proposals rather than criticisms, rather than trying to throw obstacles in the way of something that the federal government is trying to do" (U.S. Congress, House Ways and Means Committee, Hearings on H.R. 6385: Taxation of Marihuana, 75th Cong., 1st sess., Apr. 27, 1937, 116-17).

404 *81 Congressional Record* (1937), p. 5575.

405 Booth, p. 188.

406 And also influenced. Booth (p. 188) notes that "when the Senate committee debated the [Marihuana Tax Stamp Act], its passage was given a smooth ride by the chairman who was a close friend of DuPont." The idea was not so radical at the time; seventeen years earlier Mexico had passed its own federal anti-cannabis legislation, which historian Isaac Campos (p. 182) says "demonstrated how drugs were helping to justify a new, more centralized approach to governance." Ironically, the US law which Mexican drug policy inspired led to an explosion of marijuana cultivation in Mexico, which became the primary supplier of the American demand for illegal cannabis (Campos p. 226).

407 "Passage of the Marijuana Tax Act in 1937 placed all *Cannabis* under the control of U.S. Treasury regulations due to fears of the plant's psychoactive properties. This effectively prohibited the cultivation of hemp in the U.S." (Fortenbery and Bennett, p. 7). Booth, p. 188: "The Act might have been aimed at the non-existent drug fiend, but it hit hardest at the legitimate hemp-farming industry."

408 The nation's pharmacists, too, immediately ceased stocking hashish; as one J.T. Huffman of Manito, Illinois, remarked, "I have decided to discontinue the collection and sale of the herb owing to the fact that it has been placed in the narcotic list by both State and Federal laws. I have no cannabis… and have not made any collection this season as practically all of the manufacturers and dealers whom I have done business with have decided to discontinue the use and sale of this herb" (quoted in Sloman, p. 102).

409 Werner, p. 66.

410 Booth, p. 189: "With Schlicten's decorticator, which the inventor had been refining and improving ever since he registered the patent in 1919, hemp could

have made the US self-sufficient in natural fibres and reduced, if not removed, the reliance upon imported fabric and thread."

[411] Bourrie, pp. 12-13; Small and Marcus, p. 316.

[412] Booth, pp. 190-91.

[413] See Sloman, pp. 103-04; Bourrie, pp. 50-51.

[414] Grinspoon, pp. 12-13.

[415] Small and Marcus, p. 1; Booth, pp. 192-93; Bourrie, p. 55; Fortenbery and Bennett, p. 7; Illinois Industrial Hemp Investigative and Advisory Task Force Report (2000).

[416] Ironically, the Nazi government instituted a similar program to encourage Germans to grow more hemp (Small and Marcus, p. 1; Booth, p. 193).

[417] According to Campos (pp. 2, 83-84, 173-177), it was actually a Mexican medical student who first proposed the theory in 1886. He also notes that "there was considerable incentive among ordinary criminals to play up the effects that marijuana and other drugs had on them" (p. 176) potentially reinforcing the popular connection between cannabis smoking and violent insanity.

[418] Booth, p. 183; Abel, p. 159.

[419] Booth, p. 191; Sloman, pp. 111-13; Bourrie, p. 45. Munch later reported of his trial of marijuana that "I was curious, being a pharmacologist, and thought I'd try it once and see what it was all about before I let anybody else fool with it, and I found out that… well, you got the picture of what happened to me there." Western medicine has a long tradition of doctors and pharmacologists experimenting with drugs on themselves to observe the drug's effects (see V.S. Robinson, *An Essay on Hashish* [1912]).

[420] Booth, p. 192; Bonnie and Whitebread, pp. 178-79; Sloman, pp. 113-14. Anslinger wrote the doctor, "We have reliable information that the word is being passed along through the young underworld to 'blame it on the weed' when tried for a crime."

Made in the USA
Lexington, KY
18 September 2014